P9-DII-691

SAGE LIBRARY OF SOCIAL RESEARCH 39

ROBERT L. LINEBERRY

Equality and Urban Policy

The Distribution of Municipal Public Services

EQUALITY AND URBAN POLICY

Volume 39, Sage Library of Social Research

SAGE LIBRARY OF SOCIAL RESEARCH

HD 4605
L55

Equality and Urban Policy

The Distribution of Municipal Public Services

ROBERT L. LINEBERRY

VOLUME 39
SAGE LIBRARY OF
SOCIAL RESEARCH

NOV 21 1977

 SAGE PUBLICATIONS   Beverly Hills   London

Copyright © 1977 by Sage Publications, Inc.

All rights reserved. No part of this book may be reproduced or utilized in any form or by any means, electronic or mechanical, including photocopying, recording, or by any information storage and retrieval system, without permission in writing from the publisher.

For information address:

SAGE PUBLICATIONS, INC.
275 South Beverly Drive
Beverly Hills, California 90212

SAGE PUBLICATIONS LTD
St George's House/44 Hatton Garden
London EC1N 8ER

Printed in the United States of America

Library of Congress Cataloging in Publication Data

Lineberry, Robert L
    Equality and urban policy.

    (Sage library of social research; v. 39)
    Bibliography: p. 199
    Includes index.
    1. Municipal services—United States. I. Title
HD4605.L55   352'.073   76-53962
ISBN 0-8039-0742-7
ISBN 0-8039-0743-5 pbk.

FIRST PRINTING

CONTENTS

PREFACE

This book is concerned with what I take to be a fundamental, but much-neglected, problem in the study of urban politics: What city governments do with their money and who gets the benefits of urban policy. While the focus is on a single city, the issues raised are general, and only a small part of the book is consumed analyzing case-study data. The problems I discuss—service allocation and discrimination, the relationship between power structure and policy, the legal and operative standards for equality in urban policy, and bureaucratic monopoly in service delivery—are generalized issues in city politics. I hope that the effort here will constitute an opening of inquiry beyond the analysis of expenditures alone, and toward their distribution. There is more to urban policy than a budget. Behind that budget lurks the people who spend it and the consumers who receive its purchases of goods and services.

I am indebted to more people than I can name, particularly keepers of urban data who almost universally made them available. In particular, though, for advice, counsel, or assistance at various stages, I am indebted to Steve Yeager, Susan Knight, Rod Welch, Jane Macon, Angela Agee, and Patsy Taylor. The National Science Foundation provided financial assistance for the collection and analysis of data reported herein, but naturally bears no responsibility for my interpretation. Both my wife Nita, and Rhoda Blecker at Sage have exhibited the supreme virtue of patience.

October 6, 1976
Evanston   R.L.L.

Chapter 1

PUBLIC POLICY, EQUALITY, AND THE

PROBLEM OF DISTRIBUTION

Harlem was getting fucked over by everybody, the politicians, the police, the businessmen, everybody. . . . We'd laugh about when the big snowstorms came, they'd have the snowplows out downtown as soon as it stopped, but they'd let it pile up for weeks in Harlem.

Claude Brown[1]

In the section of Roxbury in which I live we have been fighting for street lights for quite some time. But they have completely ignored us. Our street is dark and though we have been writing letters and we have been getting some answers, nothing has happened. I feel it is because this area is predominantly Negro. If it was any other area they would have gotten action.

Mrs. Merle Springer[2]

Seldom specified is how goods and services get distributed to population groups.

John Patrick Crecine[3]

In the tiny town of Shaw, Mississippi, 97 percent of the homes without sanitary sewers were black dwelling units; 98 percent of the town's houses not fronting on paved streets were black-occupied; all the city's new mercury vapor street lighting went to all-white neighborhoods. Judge Elbert Tuttle, responding to the claims of black citizens in "Hawkins v. Shaw," writes that

Referring to a portion of a town or a segment of society as being "on the wrong side of the tracks" has for too long been a familiar expression to most Americans. Such a phrase immediately conjures up an area characterized by poor housing, overcrowded conditions, and, in short, overall deterioration. While there may be many reasons why such areas exist in nearly all our cities, one reason that cannot be accepted is the discriminatory provision of municipal services based on race.[4]

The relationship between the urban public sector and the "other side of the tracks" is the subject of this study. However massive the federal budget, most governmental activities, especially those immediately affecting the lives and fortunes of citizens, are provided by urban governments outspend the federal government on domestic programs. The services performed by municipalities are those most vital to the preservation of life (police, fire, sanitation, public health), liberty (police, courts, prosecutors), property (zoning, planning, taxing), and public enlightenment (schools, libraries). Michael B. Tietz, a regional scientist, emphasized the importance of urban public services to the consumer's life cycle:

> Modern urban man is born in a publicly financed hospital, receives his education in a publicly supported school and university, spends a good part of his time travelling on publicly built transportation facilities, communicates through the post office or the quasi-public telephone system, drinks his public drinking water, disposes of his garbage through the public removal system, reads his public library books, picnics in his public parks, is protected by public police, fire, and health systems; eventually he dies, again in a hospital, and may even be buried in a public cemetary [sic]. Ideological conservatives notwithstanding, his everyday life is inextricably bound up with government decisions on these and numerous other local services.[5]

How such services are allocated within the city is one of the most enduring questions of urban governance, and one which we investigate in the present volume.

The urban mosaic represents a sociospatial distribution of countless groups occupying relatively limited space.[6] That individuals are not randomly scattered over the metropolitan landscape is the first fact of urban life. Both the private and the public sector contribute to this sorting out of identifiable groups in spatial form. The distribution of wealth, patterns of racial and economic discrimination, access to jobs and housing, real estate practices, and a host of other variables all produce a more or less involuntary sifting of the metropolitan population into distinctive sociospatial groupings. On the public side, zoning practices, suburban incorporation, the location of public housing, transportation networks, and a host of other governmental decisions reinforce, alter, or reverse the location decisions of persons and production. Our concern is not to explain the public and private forces which sort different types of people into different "turfs." We take this almost as a "given."

Our interest, rather, is in the variability—if any— in the packages of public goods and services allocated to sociospatial groups within the city. We explore Lasswell's aphorism: Who gets what how. It is a common belief that all do not share equally in the bounties of public policy. The first two quotations at the beginning of this chapter typify this belief. From schools to sewers, qualities and quantities of services may be highly variable. One area within the city may enjoy services commensurate with the best that administrative knowhow and tax dollars can provide. Other neighborhoods, whether for reasons of historical accident, overt discrimination, or whatever, may constitute the wrong side of the public tracks, containing the poorest schools, the least-swept streets, the oldest and most deteriorating parks, the slowest police protection, the most potholes, and so forth. Sometimes such disadvantages may be specifically racial in character, as when the county road grader in one southern county picked up its blade in black areas.[7] Possibly such disadvantages may be highly correlated with the socioeconomic attributes of a particular area, or with the age and density of a neighborhood, or with its political power in city hall.

This volume addresses not only the general issues of public policy and the allocation of funds in the urban context, but also reports a study of public services within one city, San Antonio, Texas, the nation's eleventh largest city. It is our hope to provide data where there is now mostly discussion, policy analysis where there is now mostly polemics. The analysis of urban services may be described as a parameteorological phenomenon: There is a lot of talk, but few have done anything about it. We investigate the burdens and the benefits of urban policy. Crucial are both questions of *whether* and questions of *why*. First it is essential to

establish whether service patterns are equal or unequal, discriminatory or nondiscriminatory, redistributive or not. Then we must
examine—within the causal confines of current social science
methods—why one or another distributional pattern prevails. If
services are unequally distributed, the most common suspicion is a
variant of the "underclass" hypothesis, i.e., that some groups
suffer because of their race, because of their social status, or
because of their paucity of political power. But there are other
possible explanations for the distribution of urban services. One
holds that ecological attributes of neighborhoods—their age,
density, geography, demography, and so on—contribute to the
character of the services they receive. A final hypothesis contends
that urban service distributions are a straightforward function of
bureaucratic decision-rules. The criteria for allocating public
burdens and benefits are made by and for the urban bureaucracy,
and their discrimination, if any, may be an incidental externality
of rules made for "professional" reasons. Obviously, these explanations are not mutually exclusive, but they are amenable to
empirical assessment. We investigate in Chapters 3-5 the actual
patterns of service distribution in San Antonio, seeking to determine not only whether services are unequal, but why a particular
pattern prevails.

The Quiet Revolution in Urban Services

THE POLITICS OF PUBLIC SERVICES

Public services are the grist of urban politics. It is remarkable,
therefore, how little attention has been paid to their delivery and
allocation until very recently. For a long time, it was distinctly
unfashionable for political scientists to sully their hands with
questions of public services. The entire field of public administration was written off as "counting manhole covers." What
passed for analysis of urban politics was largely the study of
community power structures, which supposedly made the "big
decisions" regarding the city. This was gradually displaced by the
urban crisis literature, which described the state of the cities in
dramatic, but very general rhetoric. Today, however, such debates
have a hollow ring. Douglas Yates remarks that

after a decade of protest and demands for participation and community control, urban government appears to be entering a new era. Now that the "urban crisis" has been discovered, debated, and in some quarters dismissed, government officials and academic analysts alike have increasingly come to focus on "service delivery" as the central issue and problem of urban policy making.[8]

It is, as Yates observes, "difficult to see how a government can solve its dramatic problems if it cannot solve its routine ones." Public services—from crime prevention to street cleaning to schools—are the principal responsibility of urban governments. They consume collectively more tax dollars than any public function except national defense and are more likely than other governmental outputs to represent, quite literally, life and death matters.

Virtually all the rawest nerves of urban political life are touched by the distribution of urban service burdens and benefits. One source of the demand for community control, for example, was the dissatisfaction with the service delivery systems in minority neighborhoods. Suburbanization is allegedly hastened by urbanites' discontent with schools, police protection, and taxes. The well-known Tiebout hypothesis holds that people move about the metropolis in search of "optimal" packages of urban services and taxes.[9] Such exurbanites would rather switch than fight. Others who cannot, for economic or racial reasons, switch, may stay to fight. According to the Kerner commission, one principal cause of the racial disorders of the 1960s was dissatisfaction with municipal governments and their outputs. Among the major "grievances" of riot area residents, it ranked "police practices," "inadequate education," "poor recreation services," and "inadequacy of municipal service," as 1, 4, 5, and 10 among a list of 12 grievances.[10] In sum, the demands for desirable policy packages (good schools, neighborhood preservation through zoning, responsive police protection, better parks, and the like) and resistance to negatively valued outputs (location of sewerage treatment plants, an unwanted public housing project, and the like) run the gamut of urban conflict.

Oliver Williams and others have emphasized that urban politics is essentially a politics of spatial allocation of advantages and disadvantages. Access to strategic values is the gist of urban conflict. Williams suggests that "the teenage gang is the primitive urban political formation."[11] Gangs create and defend a turf against outsiders, often through violence but more commonly through nonviolent strategies. In the "adult" urban system, the same process occurs, though the legitimate authority of public policy displaces the crude techniques of the teenage gang. These public policy choices, from zoning to amenities, represent the use of urban services as magnets to attract or repel particular population groupings, thus distributing and redistributing the population within the metropolitan area.

What is important is to see public service decisions not merely as *objects of political conflict,* but as *fundamentally redistributive mechanisms.* They constitute, in Miller and Roby's suggestive phrase, "hidden multipliers of income." [12] To those advantaged, services represent an increment of real, as opposed to pecuniary, income; to those in receipt of less than their share, service inadequacies not only diminish real income, but symbolically emblazon their subordinate status. According to the Kerner commission, service inequalities "often take on personal and symbolic significance transcending the immediate consequences of the event. . . . Inadequate sanitation services are viewed by many ghetto residents not merely as instances of poor public service but as manifestations of racial discrimination."[13] People relate to municipal governments in small, seemingly insignificant, but cumulative ways. Citizens perform mental factor analyses, but with extremely sparse data.

Consequently, the most significant single issue regarding municipal services—and the principal focus of this book—is their allocation to sociospatial groupings within the city. The geographer David Smith observes that

> space creates inequalities. The location of every new facility favours or disfavours those nearby, and thus redistributes well-being or ill-being.

Any development of land has similar effects. How people in different areas establish differential claims on society's resources depends upon the spatial exercise of political power—a matter deserving of more attention than it usually gets in so-called "political geography". . . . Ultimately, who gets what *where* and how must be viewed as a question of equity or fairness. [14]

An analysis of urban service allocations is an inquiry into discrimination and inequality. These legal issues of discrimination have been more commonly raised regarding schools and school finance than regarding conventional urban services. In the school finance cases, most notably "Serrano" and "Rodriguez," attacks were made on school finance systems which discriminated against property-poor districts. A few cases have dealt explicitly with discrimination in the allocation of school finances within a particular school district. In "Hobson v. Hansen," Judge Skelly Wright was persuaded that the Equal Protection Clause extended to intradistrict distribution of school expenditures in Washington, D.C. [15] He ordered equalization of expenditures to ±5 percent from school to school.

The legal movement on the conventional services front has been a much quieter revolution, but one which promises greater success than the ultimately unsuccessful "Rodriquez" appeal. The constitution is itself silent on the question of urban services. But both common law and the Equal Protection Clause have opened the door to attacks on service inequalities. More recently, a "sleeper" clause in the revenue sharing legislation, one which prohibits discrimination in the use of shared funds, has provided additional leverage to challenge unequal services. Because revenue sharing funds are "fungible," meaning that their ultimate use is almost impossible to trace, the nondiscrimination clause is tantamount to forbidding discrimination in virtually any activity supported by general municipal budgets. The Lawyers' Committee on Civil Rights under Law has used both the "Hawkins" precedent and the revenue sharing provisions to attach Shaw-like denials of services to minority neighborhoods in southern communities. Federal courts have escrowed revenue-sharing monies in several Mississippi communities, pending judicial resolution of service discrimination cases. Such a weapon in the hands of vigorous legal action groups makes communities large and small, north and south, under threat to put their delivery systems in constitutional order.

Yet, while lawyers have busily taken municipalities to court over denials of inter-neighborhood service equity, social scientists have been discovering that the pattern of service allocations is much more complex than originally hypothesized. For a long time, conventional wisdom espoused an "underclass hypothesis," which we shall explore subsequently in this book. The underclass hypothesis assumed that the relationship between neighborhood income and neighborhood services was simply high-high and low-low. Howard Hallman, reporting to a Senate committee investigating the war on poverty, remarks that "municipal housekeeping departments have a double standard of service, with quality correlated to neighborhood income," though he cited no evidence to that effect. [16] Ira Lowry, a housing specialist with RAND, suggests that "there is a great deal of anecdotal evidence that low-income neighborhoods within most municipalities are less well served by municipal governments than high-income neighborhoods," but that "little systematic evidence has ever been collected on the subject."[17]

Evidence has accumulated, however, to suggest that patterns of service delivery may be highly variable from city to city and from service to service. The patterns in Syracuse, Sacramento, and San Antonio may differ from one another, and especially from those in Shaw, Mississippi. The empirical world, as social scientists have been discovering for years, is far more complex than the primitive social science hypotheses suggested. Precisely because the law is a blunt instrument, it is critical to understand that service measurement is both a normative and methodological question of considerable complexity.

Levy, Meltsner, and Wildavsky, for example, examined carefully "who got what" in Oakland, California, particularly in the service areas of schools, streets, and sewers. They drew this conclusion:

> There is an adage that the rich get richer and the poor get poorer, but in our work we found a distribution pattern that favored both extremes. Some mechanisms were biased toward the rich. Other mechanisms favored the poor. We discovered no mechanisms that favor the middle.[18]

School resources in Oakland were allocated in a fashion which favored both rich and poor against the middle-income neighborhoods. Street-maintenance expenditures tended to advantage upper-income neighborhoods. Library allocations were more complex. The decisions of the library administration in Oakland resulted in overinvestment in the main library compared with its circulation (it got 60 percent of the resources with only 25 percent of the circulation). Users at branch libraries were thus at a disadvantage, and users in poor neighborhoods were served by the smallest collections. We shall explore, in Chapters 4 and 5, both the burdens and the benefits of urban government in another large city, San Antonio, Texas.

URBAN SERVICES AND PUBLIC POLICY-MAKING IN THE CITY

Identifying patterns of service delivery does not, however, explain them. Such patterns may be conscious or serendipitous, intended or unintended. If discriminatory patterns are discovered, it will not suffice to attribute them to evil motives of "institutional racism." Service decisions are the product of the urban policy-making process. That process occurs within a structure composed broadly of urban elites, elected officials, interest groups, and the delivery bureaucracies of municipal governments. The power structure literature suggests implicitly that service decisions are upwardly redistributive, providing the greatest tax advantages and service largesse to the politically powerful. The countervailing pluralism literature implies that service decisions are results of a pulling and hauling of local interest groups and elected decision-makers. Fortunately, some of these hypotheses are empirically testable. In our analysis, we examine specifically the connection between power structure—in a city which contains one of the most tightly-knit power structures in the nation—and patterns of service allocation.

In addition, the analysis of urban services has led to a rediscovery of that terra incognita of urban government, the municipal bureaucracy. Jones, Levy, Mladenka, and others who have examined urban services have stressed the dominant role which bureaucratic decision-rules play in allocations. [19] Bureaucracies develop decision-rules in considerable autonomy from external elements and apply rules to routinize the allocation decisions they make. Such rules are not explicitly but implicitly distributional. One might even suggest that the allocation of urban services is a function of the cumulation of minor, routine decisions. Decisions made overtly for bureaucratic convenience nonetheless have non-neutral implications for Lasswell's aphorism about who gets what.

As in almost all things urban, New York City is either sui generis or a foretaste of things to come. The role of New York's massive urban bureaucracies in securing not only wage benefits from the city, but substantial control over service decisions, is well known. The sharpest attacks on the "monopoly bureaucracy" come also from New York City. Elsewhere, bureaucracies operate more quietly, but with the same dominance of the routine decisions. The growing impotence of elected officials (and, some sociologists to the contrary, urban power structures) is paralleled by the escalating control of public sector hierarchies. We shall trace the role of bureaucratic decision-making and the allocation of public services in Chapter 6.

URBAN SERVICES AND THE POLICY APPROACH

The analysis of "public policy" has become a major growth stock in the social sciences. The legion of scholars who fly the policy flag is too disparate to be readily catalogued. Two segments, though, of the policy community will find the analysis of urban services useful to theoretical development.

The group of scholars, largely in political science, who have focused on "policy outputs" of states and localities have utilized statistical models to identify the principal sources of variation in state and local policy.[20] Dye, Sharkansky, Hofferbert, and others have focused on policy outputs at the state level, while others have done similar research at the local level. At the same time, students of state and local budgeting examined the highly predictable, incremental patterns of budget-making. The state and local policy analysts typically discovered that "economics counts more than politics" in predicting variation in expenditures. The budgeting literature also tended to discount political factors, finding that budgets were products of internal bureaucratic logics, pretty much beyond the politician's control. Political factors—in the conventional sense of voters, councils, turnout, parties, groups, legislators, and the like—seemed to be as inconsequential in explaining budget-making as in explaining policy outputs.

The principal weaknesses, however, in these approaches have been observed as often by the practitioners as by the critics. First, there has been a fixation on the fiscal side of policy. Expenditures and taxation are the most common indicia of policy. Thus, even while some members of the policy analysis school were busily arguing that expenditures were poor surrogates for services delivered, [21] the rest persisted in taking spending measures as the most important indicator of state and local policy. Secondly, most policy research has been concerned with the *level* of expenditure rather than with its distribution. Yet, there is no necessary relationship between the size of the public pie and its distribution. Lasswell exhorted political scientists to concern themselves with who got what; the political analysis of state and local policy has more commonly emphasized "who spends how much."

One leader in the analysis of state policy summed up the principal weakness of the research stream when he observed that "perhaps the most serious reservation regarding this research is its failure to examine distributive and redistributive aspects of state [and local] politics." [22] Martin Katzman makes an observation about economists who deal with public finance:

> The empirical study of distribution as a key economic process seems relatively neglected in the professional literature. This is especially true for distribution in a nonmarket contest. While different resource allocations produce different income distributions, studies of allocation in the public sector often ignore distributional aspects as if some lump sum transfer could painlessly solve the problem of equity. [23]

If distribution is an important economic process, it is also an important political one. If we turn our attention from the levels of policy, and cease defining public policy as a figure preceded by a dollar sign, we should be better able to say something meaningful about public policy as the "authoritative allocation of values." [24]

There is at least one other major sector of the public policy school which is concerned about urban governments and their services. These are the "policy scientists," who, armed with a battery of analytic techniques, attempt to mediate the social technology function with the political decision-making function.

Policy sciences first emerged in the fields of natural resources management and in the Department of Defense. The cost-benefit analysis was the staple commodity of the earlier efforts. More recently, a significant issue in policy analysis has been the search for productivity improvements and measurement in the public sector. The attempt to get more bang for the buck began first with the most financially hard-pressed cities—notably New York City— and subsequently spread as financial crises impended elsewhere. [25] Questions related to productivity, though, are not unrelated to questions of benefits. Improvements in efficiency can easily be purchased at the price of reductions in equity. Knowing *how* a parks department allocates its resources is a crucial part of understanding *what* it is producing. A production function without allocational implications represents a most incomplete statement of the performance of the urban public sector.

An Overview

The 1.7 million nonschool employees of local governments in the United States produce public services whose value (as opposed to their cost) cannot be even roughly measured. They clearly involve the core values of life (police and fire protection), liberty (law enforcement), public enlightenment (libraries and schools), wealth (zoning and taxing), and public happiness (parks and recreation). We take Lasswell's definition of politics as our starting point, by examining *who gets what how,* adding a perspective from the political geographer by focusing as well on *where.*

In Chapter 2, we examine the problem of equality in normative theory and in constitutional law and examine its implications for the analysis of urban public services.

Chapter 3 takes an empirical approach to the subject of urban services. There, a methodology for the analysis of distributional decisions is outlined and a case in point is introduced. Chapters 4 and 5 represent the fruits of this methodology. In Chapter 4, we examine the burdens of urban policy, focusing on the question of tax assessments and the uneven burden of public finance. In Chapter 5, the distribution of services is explored, and we set about testing the relationship between race, class and power (which collectively constitute the "underclass hypothesis"), and the allocation of public services.

Chapter 6 moves back from the particular case in point to a more general issue—the decisional process surrounding urban services and its implications for the problem of equality in the city. The final chapter draws together these methodological, constitutional, normative, and empirical strands by setting forth "The Limits to Equality" as a distributive principle for urban public service allocations.

NOTES

1. Claude Brown, *Manchild in the Promised Land* (New York: Macmillan, 1965), p. 193.

2. United States Commission on Civil Rights, *A Time to Listen, A Time to Act* (Washington, D.C.: U.S. Government Printing Office, 1967), pp. 18-19.

3. John Patrick Crecine, ed., *Financing the Metropolis* (Beverly Hills: Sage, 1970), p. 515.

4. *Hawkins v. Shaw,* 437 F. 2d 1286, 1287 (5th Cir. 1971).

5. Michael B. Tietz, "Toward a Theory of Urban Public Facility Location," *Papers of the Regional Science Association,* 11 (1967), p. 36.

6. For an exposition of the sociospatial theory of urban politics, and the importance of location choices therein, see Oliver P. Williams, *Metropolitan Political Analysis* (New York: Free Press, 1971), and Kevin R. Cox, *Conflict, Power, and Politics in the City: A Geographic View* (New York: McGraw-Hill, 1973).

7. William R. Keech, *The Impact of Negro Voting* (Chicago: Rand McNally, 1968), p. 98.

8. "Service Delivery and the Urban Political Order," in Willis D. Hawley and David Rogers, eds., *Improving the Quality of Urban Management* (Beverly Hills: Sage, 1974), pp. 213-214.

9. Charles M. Tiebout, "A Pure Theory of Local Expenditures," *Journal of Political Economy,* 64 (October, 1956), 416-424.

10. National Advisory Commission on Civil Disorders, *Report* (Washington, D.C.: U.S. Government Printing Office, 1967), pp. 80-83.

11. Williams, *op. cit.,* p. 43.

12. S. M. Miller and Pamela Roby, *The Future of Inequality* (New York: Basic Books, 1970), ch. 5.

13. National Advisory Commission on Civil Disorders, *op. cit.,* p. 148.

14. David M. Smith, "Who Gets What *Where* and How: A Welfare Focus for Human Geography," *Geography,* 59 (November, 1974), p. 294.

15. *Hobson v. Hansen,* 269 F. Supp. 401 (D.D.C. 1967), aff'd sub nom. *Smuck v. Hobson,* 408 F. 2nd 175 (D.C. Cir. 1969).

16. Howard W. Hallman, "The Community Action Program," in Warner Bloomberg and Henry Schmandt, eds., *Power, Poverty, and Urban Policy* (Beverly Hills: Sage, 1968), p. 309.

17. Ira S. Lowry, "Housing," in Anthony H. Pascal, ed., *Cities in Trouble: An Agenda for Urban Research* (Santa Monica, Calif.: Rand Corporation, 1968), p. 22.

18. Frank S. Levy, Arnold J. Meltsner, and Aaron Wildavsky, *Urban Outcomes: Schools, Streets, and Libraries* (Berkeley: Univ. of California Press, 1974), p. 219.

19. Bryan D. Jones, et al., "Bureaucratic Response to Citizen Initiated Contacts: Environmental Enforcement in Detroit," *American Political Science Review* (forthcoming); Levy, et al., *op. cit.;* Kenneth Mladenka, "Serving the Public—The Provision of Municipal Goods and Services," unpublished Ph.D. dissertation, Department of Political Science, Rice University, 1975.

20. For a review of these studies, see Richard I. Hofferbert, "State and Community Policy Studies: A Review of Comparative Input-Output Analyses," in James A. Robin-

son, ed., *Political Science Annual,* vol. 3 (Indianapolis, Ind.: Bobbs-Merrill, 1972), pp. 3-72. On budgeting policy, see John P. Crecine, *Governmental Problem-Solving* (Chicago: Rand McNally, 1969).

21. Ira Sharkansky, *The Routines of Politics* (New York: Van Nostrand Reinhold, 1970), pp. 106-123.

22. Thomas R. Dye, "Income Inequality and American State Politics," *American Political Science Review,* 63 (March, 1969), p. 157. See, however, the distributional analysis of state policy outputs in Brian R. Fry and Richard F. Winters, "The Politics of Redistribution," *American Political Science Review,* 64 (June, 1970), pp. 508-522.

23. Martin T. Katzman, "Distribution and Production in a Big City School System," *Yale Economic Essays,* 8 (September, 1968), p. 201.

24. The term, of course, is David Easton's widely cited definition of politics, found in his *The Political System* (New York: Alfred A. Knopf, 1953).

25. On productivity analysis in urban government, see John P. Ross and Jesse Burkhead, *Productivity in the Local Government Sector* (Lexington, Mass.: D. C. Heath, 1974).

Chapter 2

EQUALITY, PUBLIC LAW, AND PUBLIC SERVICES

The city has been under siege, both by those within and those without, to live up to an equality norm in the treatment of its inhabitants. Often, of course, the demand is made by people who are themselves quite unprepared to live with the consequences of their demands being met.

Norton Long[1]

Once loosed, the idea of equality is not easily cabined.

Archibald Cox[2]

Some years back, John Kenneth Galbraith observed that "inequality has ceased to preoccupy men's minds."[3] The accuracy of his observation is now questionable, unless one takes quite literally the phrase "preoccupy men's minds." Galbraith to the contrary, the issue of equality has evoked profound debate, not only in the policy domains of schools and race, its long-standing terrain, but in newer areas as well. The debate over equality has extended to problems in the biological allocation of intelligence, to political dilemmas in compensating for past inequalities, and to the most sacrosanct issue of a capitalist system, income distribution itself. *Business Week,* in a series of articles on egalitarianism, was moved to remark that "distribution of income is the central issue in the insistent push for greater social and economic equality in the U.S."[4]

This resuscitation of the equality precept has been furthered by the corresponding revival of "moral philosophy," largely through the works of two contending Harvard philosophers, John Rawls and Michael Nozick.[5] Rawls' first principles of a just social order

[25]

demand equality of distribution, unless it can be shown that inequalities would rebound to the benefit of the most disadvantaged. Nozick's classical liberal counterpoint to Rawls stresses the rights of individuals to the fruits of their own labor. Whether such debates will inform the political process cannot be foretold, but they at least permit the issue of inequality to be addressed with philosophical armor regardless of one's side. Neither philosopher, unfortunately, provides many bridging principles between the normative theories and the operational issues of public policy analysis.

Although the status of both social research and moral philosophy may have shown Galbraith's "end of equality" hypothesis to have been premature, still the Equal Protection clause has not lost its constitutional cutting edge. The Warren Court seemingly found fourteenth amendment issues in virtually every question of social policy it confronted. But despite the fears of its early critics, the Burger Court has stepped back from, but by no means abandoned, the fourteenth amendment test of state action. "Less egalitarian" decisions in metropoliswide school desegragation and school financing were balanced by "more egalitarian" decisions in metropolitan housing and school busing. As yet, no specific question of public service discrimination has reached the Supreme Court. But the Supreme Court is not the only court around, and the texture of precedent used to assault service discrimination has been carefully built upon its earlier interpretations of the fourteenth amendment. Courts in this nation are historically better equipped to deal with universalistic policy choices than legislatures. Consequently, the issue of urban public services will remain more, perhaps, a juridical than a legislative one.

We shall argue that the most serious problem in the analysis of urban public services is neither empirical nor methodological. To be sure, a number of assumptions, and considerable effort, are required to assess the sociospatial distribution of urban services. The real difficulty, however, is normative. The very terms "distribution" and "allocation" imply and invoke some *principle* by which *something of value* is disbursed. The distribution of a trivial value (say, paper clips) is of no enduring interest. Similarly, the distribution of a commodity by a trivial principle (say, blueness of the eyes) is equally uninteresting, unless the trivial principle results in discrimination. Normative standards are inextricably linked with the empirical assessment of urban public services. And the most important principle, in both moral philosophy and constitutional law, which can be invoked, is equality. The importance of the principle as a test of public policy was long ago stressed by Rousseau's penetrating observation that "it is precisely because the force of things tends always to destroy equality that the force of legislation ought always to tend to maintain it."

The geographer David Harvey has made a convincing case for the proposition that "the 'hidden mechanisms' of income redistribution in a complex city system usually increase inequalities rather than reduce them."[6] If this is correct, then the presumably unseen hands which produce vast wealth also sift and sort urbanites into economic strata and identifiable sociospatial clusterings. The poor would thus suffer a kind of double jeopardy if the public sector allocated its policies to further advantage the advantaged. Thus, Rousseau's dictum holds that public policy—the "force of legislation"—should equalize precisely to counterbalance the inequalities of the marketplace. Whether it does or not is the subject of this analysis.

Yet no philosophic-legal principle—save perhaps "justice"—is less susceptible to operational treatment than "equality." It remains, as Fitzhugh Stevens said, "a word so wide and vague as to be almost nonmeaning."[7] Despite its centrality to the law, to philosophy, and to the social sciences, the concept of equality is not merely complex but fundamentally paradoxical. How well it will serve as a distributional standard depends upon how well its implications can be operationalized. Here we are concerned with its meanings for public law, public opinion, and public policy.

The Idea of Equality

There is some dispute about its antiquity, but equality is not a recent notion. The *Digest of Roman Law* even held that "according to the natural law, all men are equal."[8] It was also a crucial concept for the contract theorists, who took it to be the "natural" condition of pre-civil society. Montesquieu argued, in a prescient passage, that "in the state of nature, indeed, all men are born equal, but they cannot continue in this equality. Society makes them lose it, and they can recover it only by the protection of the laws."[9] Today, no idea ramifies throughout as many issues, from schooling to suffrage, from race to reapportionment, as does equality. The fourteenth amendment's equal protection clause portends as many constitutional clashes as any element of the legal system.

The simplicity of the concept is its principal virtue and its main defect. Patent inequalities affront the consciences of courts and citizens in a value system infused with egalitarian premises. When Shaw, Mississippi, fails to provide sewers to black neighborhoods while doing so for whites, when San Francisco provides the 7 percent of its population residing in Chinatown with 1.41 acres of its 5,282 acres of parks, equality becomes a useful moral (if not always a legal) weapon.[10] Yet, as the examples indicate, most contemporary discussions of equality do not propose equality directly, but rather dispute about the tolerableness of inequalities. Even the fourteenth amendment is phrased negatively: States are not compelled to provide equal protection of the laws, but rather forbidden to deny it. Most polemics do not support actual equality of condition as much as they rail against extreme inequalities. Demands for income *redistribution*, though defended with egalitarian principles, are still a far cry from income *equality*. In fact, equality is almost never taken literally as a guide to social policy. No egalitarian envisages the obliteration of distinctions among persons. "To make equality a symbol for the absence of all distinctions," says Lakoff, "is not to define any of the real proposals of equality, but only to prepare an attack on absurd caricatures of them all."[11] Egalitarianism is thus selective. But if the concept is not to be over-literalized, it becomes essential to define the objects of equality. The serious proponent prescribes equality only with respect to his dominant social values. Thus, the Levellers advocated redistribution of land, while in contemporary western society, egalitarians are more concerned with civil liberties or with wealth and income.

Obviously, then, if equality is to apply only to *some* values (or public policies), the problem is to specify those core values as objects of equality. Consequently, "arguments about equality are, thus, arguments about criteria of relevance."[12] The Warren Court attempted to resolve the problem by identifying certain "fundamental interests" which were to be inviolable against government action. At one time or another these included the right to vote, to travel interstate, to join political associations, as well as certain procedural rights as the bar of justice. That the Burger Court has "declassified" some of these does not lessen the problem of specifying the objects of equality. Whether municipal governments possess powers fundamental enough to amount to an egalitarian mandate is a significant question. In any hypothetical hierarchy of municipal services, a line can be drawn somewhere between those deemed essential and those less crucial to the preservation of citizen equality.

Equality is a complex and uncertain guide to public policy, though, not merely because its objects are as numerous as its proponents, but also because (paradoxically) equality may promote inequalities while inequality may promote equality. The long-run pursuit of equality may tolerate, even require, shortrun inequalities of a "compensatory" nature. Alternatively, inequalities with respect to one value may be tolerated because they contribute to maximization of equality on a higher value. Compensatory education is anything but equal education, although equality is its rationale. In "Griffin v. Illinois," the Supreme Court required the state to eliminate the effect of a private-sector inequality, indigence, by substituting a public-sector inequality, free court transcripts. Dissenting, Justice Harlan grumbled his objections to the paradoxical character of the majority holding, fearing that the decision "imposes on the states an affirmative duty to lift the handicaps from differences in economic circumstances."[13]

The ultimate paradox of equality is the "meritocracy problem." Eloquently introduced by Michael Young's social science fiction analysis of the British aristocracy, the argument is a philosophical analogue of Arrow's impossibility theorem. The more equalizing society becomes, it holds, the more unequal are its members. As "unnatural differences" are eradicated through social reforms, "natural differences," principally differences in intellectual skills, produce a "meritocracy." The psychologist Richard Herrnstein has elevated the meritocracy hypothesis to a few simple propositions involving intelligence:

> The main significance of intelligence testing is what it says about a society built around human inequalities. The message is so clear that it can be made in the form of a syllogism:
> 1. If differences in mental abilities are inherited, and
> 2. If success requires these abilities, and
> 3. If earnings and prestige depend on success,
> 4. Then social standing (which reflects earnings and prestige) will be based to some extent on inherited differences among people.[14]

The hypothesized upshot is a meritocracy in which the natural and intellectual aristocracies become virtually coterminous. Social reform via equalizing is thus taking one step forward, then one step back.

Equality and Public Policy

Most contemporary discussions of equality relate to the schools. Coleman's *Equality of Educational Opportunity*, in particular, has become perhaps the most controversial piece of modern social science ever published.[15] If Coleman found that educational resources were relatively inefficacious in equalizing student performance, then Christopher Jencks did him one better. Jencks contended that performance in schools was not appreciably related to life chances, save as a credentialing mechanism.[16] The issue of educational equality surfaces, too, in the courts. In a series of school resource cases involving both inter-district and intra-district resource differentials, plaintiffs challenged the constitutionality of school finance mechanisms. In Texas, California, Minnesota, New Jersey, Arizona, and elsewhere, federal and state courts held against finance systems which tied local school taxes to wealth. The Supreme Court, however, held against such claims in its review of "San Antonio Independent School District v. Rodriguez."[17] Rarely, however, as we shall see, has the Supreme Court faced the issue of the distribution of noneducational public services. Yet, we can learn something from the promise and pitfalls of equality applied to schools.

INPUT AND OUTPUT EQUALITY

James Coleman first drew the distinction between *input* and *output* equality in the measurement of educational systems. There are, he suggests, five possible meanings of "equal education": (1) equal facilities and teachers; (2) racial integration; (3) intangibles, such as prestige, alumni support, morale, and so forth; (4) equal educational outcomes given *equal* pupil capacities and backgrounds; and (5) equal educational outcomes given *unequal* pupil capacities and backgrounds.[18] The first three of these represent input equality, i.e., the delivery of comparable packages of resources, whether tangible or intangible. At one time or another, the courts have used each of these criteria as tests of both inter- and intra-district denials of equal protection. Both the major school finance cases, "Serrano" in California and "Rodriguez" in Texas, rely upon standard (1) on an inter-district basis, while "Hobson v. Hansen" used it to order expenditure equalization in the District of Columbia. Standard (2) is the basis for a series of desegregation cases from "Brown v. Board of Education" to the school busing case of "Swann." Standard (3) was evident in "Sweatt v. Painter," where such intangibles as alumni prestige were considered in the Supreme Court's decision to integrate the University of Texas Law School.

But the more controversial standards are Coleman's output equality measures. However faint the chances of elevating it to a constitutional test, Coleman nonetheless maintains that

> the schools are successful only insofar as they reduce the dependence of a child's opportunities upon his social origins. . . . *Thus equality of educational opportunity implies not merely equal schools, but equally effective schools,* whose influences will overcome the differences in starting points of children from different social groups.[19]

If, as Coleman found, initial pupil disparities actually widen over twelve years, then schools flunk the test of output equality.

However appealing, the standard of output equality is also politically controversial, judicially improbable, and educationally difficult to guarantee. It does, though, point up the limits to input equality. The great danger in an input equality test is that it will be taken too literally, giving rise to Lakoff's fear of an "absurd caricature" of a complex concept. Nowhere is this problem more apparent than in the zealous pursuit of one input dimension, resources, to the exclusion of other tests. Expenditures are, unfortunately, a problematic guide to quality. Sharkansky, for example, discovered only weak to moderate correlations between state expenditures and measures of state service outputs.[20] Dollar increases in public spending may reflect the purchase of higher services, or merely register increasing inefficiency. Expenditure differences may be affected by geographical variations in the costs of labor and materials, by varying rates of wear, deterioration, or weathering, by differential rates of vandalism, and by variations in the nature of the pupil composition of the district. The standard of output equality is intended to rectify these deficiencies by focusing on performance measures rather than resource measures. Output equality thus requires *equality of condition after receipt of a service,* be it education, policing, or whatever. To achieve output equality almost necessitates input inequality. To achieve input equality, police department resources must be dispersed equally over the municipality; but to achieve output equality, resources will have to be clustered in high-crime neighborhoods if victimization rates are to be equalized. This assumes, of course, that more police resources *can* reduce the crime rate, a proposition which is itself disputable.

Despite the popularity and superficial appeal of Coleman's output equality standard, there are several objections to it as a guide to service distribution. In the first place, the ultimate goals of any public service—take parks and libraries, for example—are typically so general and vague ("to provide recreation," "to enlighten the public") that the standard is indeterminate. What exactly might constitute output equality in the service areas of sanitation, parks, streets, libraries, and traffic control? Even in the realm of education, where the concept originated, the only real measure of output equality is the achievement test score. In the second place, measurement of output equality assumes the specification of causal input-output linkages. Altering the outputs of a system requires knowledge about given combinations of output-input relationships. Yet as the great debate over the impact of education illustrates, the input-output nexus is tenuous at best.

Only once has the input-output distinction been explicitly raised about the distribution of municipal services. In "Beal v. Lindsay,"[21] the plaintiffs, minority residents in New York City, demonstrated that their neighborhood park was inferior in condition to other parks in the city. The city, however, demonstrated that the park, nonetheless, received equal inputs of resources and personnel. The deteriorated condition of the park was a result, it argued successfully, of higher vandalism rates in the neighborhood. Failure to obtain output equality in recreation services did not violate the constitutional requirement of equal protection. This outcome suggests that the concept of output equality probably has a brighter future as a moral standard than a legal one. It would be extraordinarily difficult to guarantee output equality with respect to most conventional urban services. If the evidence in law enforcement is any guide to other policies, higher levels of police inputs may have only tangential effects on securing greater outputs, specifically, lower crime rates.[22]

EQUALITY, COMPENSATION, AND LEVELLING

Dispairing of the paradoxical character of equality, some have relied on "equality of opportunity." Yet that principle has been condemned by critics as a snare and delusion. However virtuous the notion that everyone should start with an equal shot at society's values, with only talents limiting success, John Schaar has asked:

> What is so generous about telling a man he can go as far as his talents will take him when his talents are meager? Imagine a foot race of one mile in which ten men compete, the rules being the same for all. Three of the competitors are forty years old, one has weak ankles, five are overweight, and one is Roger Bannister. What sense does it make to say that all ten have an equal opportunity to win the race? The outcome is predetermined by nature, and nine of the competitors will call it a mockery when they are told that all have the same chance to win.

The difficulty with "equal opportunity," Schaar concludes, is that, if followed seriously, it "removes the question of how men should be treated from the realm of human responsibility and returns it to 'nature'."[23]

But if equality of opportunity is the conservative response to the dilemmas of inequality, then compensation is the liberal answer, and levelling the radical one. Rainwater calls these latter strategies the "bottom up" and the "top down" approaches to income distribution. The bottom up approach is compensatory in that it attempts to uplift the lower strata to full participating membership in society. The top down approach exhibits a romantic appeal to radicals, although it "tends to reach a pragmatic dead end when it is discovered that distributing all of the personal income of the very rich would have a fairly small impact on the economic status of everyone else."[24]

The idea of compensation is to redress social contingencies so as to uplift the disadvantaged, insuring equality over the long run by imposing short-term inequalities. It is at this point that the compensation principle shades into the criterion of *need,* for there is no reason to compensate persons except to maximize their need-satisfaction. Government provides compensatory education funding because some children do not have their educational needs fulfilled at home; more park space is provided in densely-packed neighborhoods because the private sector does not satisfy leisure needs. Actually, there are two operational variants of the compensation principle, one weak and the other strong. They may be called *pseudo-* and *pure* compensation. The former tends merely to equalize, primarily by providing the disadvantaged with additional resources, leaving them still behind the advantaged, but not so far behind as before. Much of what passes for compensatory education tends only to increase spending in poor schools to a parity level with modal schools. Supplementing, however, is not really compensating. *Pure compensation begins only when input equality has been achieved across all units* (children, parks, schools, etc.). Everything up to that point is pseudo-compensation.

In the mighty arsenal of federal urban-oriented policies, only one—the Elementary and Secondary Education Act of 1965—mandates a pure compensatory principle. Title I of the ESEA demands that the programs funded "provide financial assistance . . . to local education agencies serving areas with concentrations of children from low-income families." Such funds were to be *in addition to,* not *in lieu of,* regular expenditures to low income schools. Consistent with the pure compensation principle, Title I guidelines demanded *equalization* (of expenditures between aided and non-aided schools), then compensation. In the language of the official HEW guidelines:

> Title I funds . . . are not to be used to supplant State and local funds which are already being expended *or which would be expended in those areas if the services in those areas were comparable to those in non-project areas.* [25]

The pure compensatory principle is articulated in the italicized clause. Unfortunately for the principle, however, the guidelines were implemented less stringently than their forceful enunciation implied.[26]

The advocacy—even the reality—of the compensatory principle is not unheard of in American politics. But, with the obvious exception of the federal (*only* the federal) income tax structure, very little rhetoric and ever fewer policies pursue equality via levelling. Levelling has long been viewed with horror by elites and masses alike in the American experience. If elites have opposed levelling on self-interest grounds, masses have feared it as warping the incentive structure or depriving a meritorious elite. Rainwater notes that "the idea of essentially equal distribution of resources does not seem attractive to most people."[27]

Even if incomes were levelled, of course, other fundamental inequalities would remain. Kurt Vonnegut tells a little social science fiction story, "Harrison Bergeron," about the supposed difficulties of equality via levelling:

> The year was 2081, and everybody was finally equal. They weren't only equal before God and the law. They were equal in every which way. Nobody was smarter than anybody else; nobody was better looking than anybody else; nobody was stronger than anybody else. All this equality was due to the 211th, 212th, and 213th Amendments to the Constitution, and to the unceasing vigilance of the agents of the United States Handicapper General.[28]

The latter's function was to equalize inequalities, not by redress, but by reduction. Harrison Bergeron was natively a genius and strong as he was bright, so he was most heavily handicapped of all.

In the race for life, "Harrison carried three hundred pounds." Here emerges the notion of compensating inequalities as an infinite regress: As successive inequalities are peeled away, new ones emerge to demand redress. So long as we can define the *objects* of equality, however, we need not fear such reductio ad absurdum.

Equality and the Analysis of Urban Services

The foregoing is intended to suggest that the concept of equality is not easily applied to any social phenomenon, most assuredly including urban public policy. An egalitarian argument will support input (resources) or output (performance) tests; it will sustain short-term inequalities of a compensatory nature based upon need; and it will tolerate a per capita equality, in which each citizen's share is proportional to any other citizen's share, regardless of need, taste, efficiency, or equity. When set alongside alternative standards, though, the picture becomes more cluttered still. If tradeoffs between equality and, say, taste, are to be permitted, which tradeoffs are to be tolerated and which forbidden? If compensatory inequalities shall be engineered to satisfy "needs," then, which needs are fundamental and which frivolous? Unfortunately, the equalitarian principle is no more settled in the law, as we shall see, than in the social sciences or philosophy. Yet, if we attempt to operate on all fronts at once in applying egalitarian principles to urban policy analysis, we shall be irretrievably lost. Perhaps we can simplify matters.

DESCRIBING DISTRIBUTIONAL PATTERNS

For simplicity, let us focus on the concept of per capita equality, which holds that, whatever measure of service output is chosen, it is to be dispersed in roughly identical proportions to each citizen-consumer. Realistically, it is almost impossible to disaggregate service outputs to the most discreet unit possible, the individual. This is not only operationally difficult, but theoretically problematic. Many services are not delivered to families or firms, but to larger sociospatial units like neighborhoods. Fixed facilities (parks, libraries, firehouses) could never be identically accessible to family units unless there were a park, library, or firehouse on every doorstep. Mobile, like fixed, services, cannot be expected to deliver identical packages to every householder. No one expects police to patrol a quiet residential cul-de-sac as frequently as a busy boulevard. But we do expect a rough equivalence of services from one neighborhood to another. The principal issue of urban service distribution, of course, is whether such inter-neighborhood equality prevails or not. The most common hypothesis is that some neighborhoods, particularly the already advantaged, get more than their share.

1. An *equal* distribution implies that all sociospatial units, regardless of their racial, ethnic, political, or economic attributes, share proportionately in service outputs. If schools, street lighting police protection, sewers, fire protection, and so forth, meet standards of proportionality, then we can label such distributions equal. This is, of course, nothing more than the concept of input equality. Exactly how we can specify performance measures to assess output equality for law enforcement, libraries, streets, and parks, is not clear at the moment. The aggregate quality of services in a city may be high or low, but still equal. Sewers may be good or bad throughout the whole municipality, but an equal pattern requires that all share alike in the inadequacy or superiority of the services.

2. A *direct* relationship between the socioeconomic characteristics of a sociospatial unit and its services implies that neighborhoods with advantages in the private sector enjoy public sector advantages as well. Regardless of the advantages inherent in the socioeconomic status, the race, or some other attribute of the residents, a direct relationship can be identified by the positive correlation between a measure of social dominance and a measure of service delivery. A direct relationship thus resembles the biblical precept (Matthew XXV, 29, RSV):

> For to everyone who has will more be given, and he will have abundance, but from him who has not, even what he has will be taken away.

More colloquially, "them that has, gets."

3. An *inverse* or *compensatory* relationship exists when disadvantaged neighborhoods secure higher quantities and/or qualities of municipal services than do advantaged neighborhoods. A

compensatory pattern is definitely not an equal pattern, as it conforms to the liberal precept of redressing imbalances in the private sector through the use of public policy. Thus, poorer neighborhoods, whose citizens can afford few books, get bigger and better libraries; because they can secure less private recreational space, they get more parks; because family resources are less munificent, they get better schools.

It will be instructive to inquire whether the conventional wisdom about service distribution is borne out by the available evidence. While the field of urban service distribution is not exactly uncharted terrain, neither has it been overstudied. Generally, there are two ways in which service outputs have been assessed. The first is simply to ask random samples of citizens about schools and services in their neighborhood as compared with the rest of the city. The second is the "hard data" approach, and relies on direct measures of service allocation. Both types are valuable and can, up to a point, compensate for the deficiencies of one another. Paradoxically, however, the two approaches have often reached different conclusions about distributional inequalities.

The Survey Approach

Thomas Hobbes laid down a rule of thumb regarding equality. "There is not," he said, "ordinarily a greater sign of the equal division of anything than that everyman is contented with his share." If this handy guide to measuring equality be valid, there is not as much inequality in urban services as commonly suspected. Rather surprisingly, most survey evidence shows urbanites believing—with one significant exception—that urban services are "share and share alike." The exception, of course, is black perceptions of poor police services in minority neighborhoods.[29] While black attitudes toward the police may be a tall tree, they should not be permitted to obscure the forest. For the most part, urban citizens tend to think that inter-neighborhood differences in service outputs are real, but neither great nor terribly significant. One comprehensive study using data collected for the Kerner commission examined attitudes toward service patterns in fifteen metropolitan areas. Overwhelming majorities of both black and white citizens (90 percent of whites and 76 percent of blacks) felt that their neighborhood get "about the same" or "better" services than other parts of the city.[30] Again, police protection was the exception to the rule.

By far the most ambitious study of citizen attitudes toward municipal services was conducted by ten of the nation's urban observatory cities (Kansas City, Missouri and Kansas, Milwaukee, Boston, Denver, Baltimore, Albuquerque, Nashville, Atlanta, and San Diego).[31] Samples of citizen-consumers were asked to compare their public services with the rest of the city. The combined response patterns, controlled first by class and then by ethnicity, are shown in Tables 2.1 and 2.2. One should rightly be suspicious of a survey instrument whose fixed choices do not include "worse" services in their own neighborhoods, but the results are still suggestive. As expected, minority groups and the poor are more likely to think that theirs is the short end of the service stick.

In light of the received wisdom that the poor are perpetually ill-served, however, the relative satisfaction of all groups, the poor included, may be more startling than the observed inter-group differences. Even when extreme groups, the very highest and the very lowest, are compared, there is overall consensus that one's own neighborhood fares fairly well. Table 2.3 does just that by

Table 2.1: Ratings of City Services in Neighborhood Compared to Rest of City, by Race*

	Race of respondent**		
Service rating	White	Spanish	Negro
Better	32.9%	27.5%	22.2%
	(707)	(39)	(104)
Same	67.1	72.5	77.8
	(1440)	(103)	(365)
	100.0%	100,0%	100.0%
	(2747)	(142)	(469)

*Data computed from 10-city file in possession of the Albuquerque Urban Observatory. Totals may not equal 100% due to rounding. The author is indebted to Professor Robert Wrinkle for making these data available.
**Orientals, American Indians and "other" excluded because of limited number of cases.

Table 2.2: Ratings of City Services in Neighborhood Compared to Rest of
City, by Class*

| Service rating | Social class of respondent | | | | | |
	Upper	Upper-Middle	Lower-middle	Upper-lower	Lower-lower	
Better	49.8%	42.6%	33.3%	25.4%	22.2%	31.1%
	(110)	(190)	(183)	(298)	(94)	(875)
Same	50.2	57.4	66.7	74.6	77.8	68.9
	(111)	(256)	(367)	(876)	(330)	(1940)
	100.0	100.0	100.0	100.0	100.0	
	(221)	(446)	(550)	(1174)	(424)	

*See note to Table 2.1.

Table 2.3: Selected Contrasts between Service Ratings of Upper and Lower-
Lower Class Groups*

| | Class | |
	Upper	Lower-lower
Courts "always" or "usually" fair	64.9%	61.2%
Crime fighting in neighborhood "better" or "same" as other areas of city	93.2	88.9
Crime fighting in neighborhood "not as good" as in other areas of city	6.8	11.1
Neighborhood schools are "better" or "same" as in other areas of city	87.8	79.6
Neighborhood schools are "not as good" as in other areas of city	12.2	20.5

*See note to Table 2.1.

examining specific services among the lowest and highest class respondents. Admittedly, social scientists have never been able to decide how big a difference must be to be substantively (as opposed to statistically) significant. But given the quantity of anecdotal evidence on service discrimination, it is surprising that the differences are no larger than they are.

The survey approach, however, contains some evident drawbacks. Asking people about reality may not always be the best test of objective reality. Many urban services are potential rather than actual in their utilization. Hopefully, fire and police services are among those public services which most people will rarely use. In a sense, the more effective some services are (e.g., police protection), the more invisible their delivery. At the same time, citizens may mistake a low crime rate for effective police protection, and vice versa. If citizens can be ill-informed about their own services, they may be doubly ignorant of other neighborhoods. Urban socializing and work experiences typically involve only one or two city neighborhoods, and extrapolating evidence to other areas may be sheer guesswork. In fact, Schuman and Gruenberg found little relationship between citizen assessments and objective indicators of service quality in fifteen cities.[32] All this is not to deny that citizen attitudes and perceptions are important. Indeed, they are crucial to creation of supportive or alienating attitudes toward city government. But such attitudes are not always concocted from the rock of hard reality. And perceptions may or may not coincide with actual deviations from an equalitarian norm.

The "Hard Data" Approach

Asking people is one way to ascertain service delivery patterns; collecting empirical data is another. Although evidence remains sparse and spotty, enough has accumulated to draw some generalizations about the sociospatial distributions of urban public services. They do not form an unambiguous pattern and exhibit differences from city to city, between one service type and another, and over time as well. Far and away the largest volume of evidence is in the domain of educational resource dispersion. Among these, Patricia Cayo Sexton's *Education and Income* is seminal.[33] She used a battery of indices of educational output, including class size, teacher qualification, facilities, and special programs. The distribution of educational services in Detroit was, in our terminology, direct: schools serving higher income neighborhoods enjoyed the best of everything, including even more free lunch and free milk programs. A good deal of subsequent research has confirmed Sexton's conclusions in other locales.[34] Yet other studies in other locations find equal or even compensatory patterns of school resources distribution.[35]

Perhaps the soundest evidence comes from studies which hold methodology constant by investigating two or more school districts with rich and exacting data. Mandel, for example, studied four school districts in Michigan (including Detroit and three suburban districts) and found variation from district to district. While Detroit (confirming Sexton's study of a decade earlier) still gave more resources to schools in well-to-do neighborhoods, the suburbs varied widely in their distributional patterns, ranging from direct to compensatory to curvilinear.[36] If any firm conclusions were to be drawn about intra-district distribution of educational resources, they would surely be heavily qualified by differences from district to district. Just as power structures may empirically range from Hunter's Atlanta to Dahl's New Haven regardless of the methodologies employed, so may service distributions range from direct to compensatory. What variables predict the score of a city, we cannot assuredly say.

The evidence on non-educational services is far more piecemeal. Much of this evidence will be reviewed subsequently in Chapters 3 and 4 when we explore data on San Antonio. For now, it will suffice to observe that evidence frequently—though not always—confirms the conventional wisdom about service patterns. To cite but an example or two, a federal district court found in Prattville, Alabama, a park system which affronts even the most relaxed definition of equality. While the white to black population distribution was one to five, the ratio of white to black park acreage was fifteen to one.[37] A prestigious study conducted under the auspices of the American Library Association investigated library services in Chicago. The author concluded that "the data in this report indicate that over the years the Chicago public library has been putting an undue share of its resources into the outlying sections of the city and not enough into the disadvantaged areas."[38] Such evidence can be (and in subsequent chapters, will be) multiplied exponentially. If municipal services are *directly* allocated in terms of the simplest definition of per capita equality, then we need not be overly concerned with more sophisticated meanings of equality. Equality can thus become the *de minimis* test of public service delivery.

Public Law and Public Services

Consistent with de Tocqueville's dictum that American political issues always become judicial ones, the issue of public service delivery has been more disputed in the courts than in the city halls and bureaucratic corridors. Though certainly not matching the contentiousness or drama of the school desegragation, reapportionment, or defendant's rights cases, the courts have entered—though certainly not rushed in—where legislative bodies have feared to tread. As Judge Skelly Wright observed in a decision mandating equalization of instructional expenditures in the Washington, D.C., schools:

> It is regrettable, of course, that in deciding this case this court must act in an area so alien to its expertise. It would be far better indeed for those great social and political problems to be resolved by other branches of government. But these are social and political problems which at times seem to defy such resolution. In such situations, under our system, the judiciary must bear a hand and accept its responsibility to assist in the solution where constitutional rights hang in the balance.[39]

There are, in fact, so many remedies at law for the citizen aggrieved by his public services that the wonder is that so few challenges have been levied at urban service outputs. Such remedies derive both from common and constitutional law. Under both criteria, there has been an explosion of concern (if not of actual cases) for equality in everything from schools to sewers to street lights to sidewalks.[40] As is sometimes the case, however, the level of litigation has failed to approximate the level of law journal commentary.

CONVENTIONAL SERVICES AND THE CANONS OF THE COMMON LAW

Municipal corporations are not required, except as state statute might dictate, to undertake to supply any particular service. But once undertaken, it must be extended with reasonable dispatch to all citizen-consumers. The antiquity of the doctrine—dating at least back to the eighteenth century—is rivalled only by its universality. In Maryland, the state court of appeals observed that when a municipality undertakes to perform the duties of a public service company, it must . . . furnish its services to all applicants within the area supplied and cannot unjustly discriminate between the consumers therein.[41]

With respect to fire protection, for example:

When a city assumes the responsibility of furnishing fire protection, then it has the duty of giving each person or property owner such protection as others within a similar area within the municipality are accorded under like circumstances.[42]

Or, in the case of public sewers:

Although there is no requirement that the city provide sewer services . . . once a city undertakes to provide a service to people in the city it must provide that service adequately and on an impartial and non-discriminatory basis.[43]

What is required of municipal corporations is also required of public utilities, whether electric, water, or whatever. The citations in defense of the non-discrimination principle in common law can be made as long as one's proverbial arm.

Indeed, most states also mandate by statute that annexing municipalities must deliver services to new areas commensurate with those enjoyed by long-time areas. (And it would be paradoxical if the law stipulated service equality as between old and new areas of a city, but ignored discrepancies among existing neighborhoods.) Many an annexation plan has foundered on the rock of inability to insure equal and adequate services to new areas. In Missouri, for example, the courts reviewing annexation proposals must determine the quality of services in the annexed area and the ability of the annexing municipality to provide quality services.[44] One might add, incidentally, that if courts now have the capacity to determine service conditions between old and new parts of a municipality, it seems reasonable to suppose that their capacities would not be strained by having to do so for assessing the intra-municipal distribution of public services.

There remains, however, a beguiling simplicity, at once both its strength and weakness, to common law service requirements. To mandate "equal and adequate" services is not, unfortunately, to define equal and adequate services. Precisely for this reason, common law is often superceded by statute. The common law precepts remain most useful to the complainant who can demonstrate blatant, direct, and unambiguous discrepancies.

ELEVATING SERVICES TO A CONSTITUTIONAL ISSUE: A BANG OR A WHIMPER?

The equal protection clause has been a restless constitutional giant. Sleeping fitfully for long years, it has been periodically awakened to flex its potentially powerful muscles in one cause or another. Long used to protect business from legislative assault, it was demolished in the Roosevelt Court era, only to rise again from its ashes to protect insular minorities from racial discrimination, and to aid, to a lesser degree, economic minorities as well.[45] In defendant rights, in suffrage, in school segregation, in reapportionment, and in other areas, the fourteenth amendment became the constitutional keystone of the Warren era.

By the latter days of the Warren Court, it appeared that three little phrases would be elevated to a constitutional test of the equal protection clause. If a *suspect classification* of persons was made by a state, or if legislation impinged upon a *fundamental interest,* then the Court would apply *rigid scrutiny* to determine its constitutionality. Each phrase was as ambiguous as the master Equal Protection clause itself, but the Court sketched in a few of the details. The issue of suspect classification was as old as the amendment itself. Actually, the word "classification" has always been misleading, implying a degree of intentionality almost never overtly present in legislation. Even in the by-gone pre-Warren era, courts would have looked askance at legislation blatantly singling out one group for favor or discrimination. The term "suspect classification" thus parallels the political scientist's use of "policy impact" as distinguished from "policy content." It requires the courts to examine, not merely the face of the legislation, but its impact on groups. But the mere showing of differential impact will not suffice to demonstrate a "suspect" classification for, as Professor Karst observes, "since most legislation has differential effects on various groups, discrimination . . . can be found in almost anything a legislature does or fails to do."[46] Classifications must, in the first place, be "reasonable," i.e., "including all who are similarly situated and none who are not."[47] But reasonableness is still only a partial test of acceptability, because "there are some traits which can never be made the basis of a constitutional classification."[48] One of these traits is race. As the Court observed, "racial classifications are constitutionally suspect . . . and subject to the most rigid scrutiny and in most circumstances irrelevant to any constitutionally acceptable legislative purposes." Such classifications must be "necessary, not merely rationally related, to the accomplishment of a permissible state policy."[49] Place of residence, to the Warren Court, also constituted a suspect classification.[50]

By the end of the Warren era, it appeared that wealth might also constitute a suspect classification. In "McDonald," the Court observed that "a careful examination is warranted where lines are drawn on the basis of wealth or race, two factors which would *independently* render a classification highly suspect and therefore demand a more exacting scrutiny."[51] As with race, the deliberateness of the discrimination is unimportant. At issue is the impact of policy on particular groups. In the words of the California Supreme Court, "none of the wealth classifications previously invalidated by the United States Supreme Court or this Court has been the product of purposeful discrimination. Instead, these prior decisions have involved unintentional classifications whose impact simply fell more heavily on the poor."[52] Still, the Warren Court was some distance from the full elevation of the wealth criterion to coequal constitutional status with the racial criterion. And new courts can always (Justice Warren in "Brown" to the contrary) turn back the clock.

When "fundamental interests" are impinged upon, courts also (under the Warren formula) give "rigid scrutiny." Unfortunately, one person's fundamental interests are another's trivial interests. The constitution is of little aid. As a result, the Warren Court's list of fundamental interests varies with the enumerator. A partial listing might include the right to vote, to associate in political groups, and to travel interstate. The court never elevated any urban policies to the hallowed list of fundamental interests, but education and parks have been defended in some pretty strong language.[53] So even in the heyday of the Warren Court, no municipal services were explicitly placed under the umbrella of "fundamental interests," and whether the Burger Court will travel where the Warren Court failed to tread is questionable.

The third of the three little phrases in the Warren equal-protection formula is "rigid scrutiny." It dates back to "Korematsu," where the Court said that "all legal restrictions which curb the civil liberties of a single racial group are immediately suspect. That is not to say that all such distinctions are unconstitutional. It is to say that the courts must subject them to the most rigid scrutiny."[54] In principle, it was always possible to find a "compelling state interest" to justify the restriction of a constitutionally protected interest. However, the Warren Court never once found a state interest sufficiently compelling to save a challenged statute. Thus the equal-protection clause was about to become, as the Warren era faded, the vehicle by which both school and nonschool urban services were equalized. Men like John Anthony Serrano, Dimitrio Rodriguez, Andrew Hawkins, and Julius Hobson challenged the distribution of urban tax and expenditure policies and, by implication, the foundations of autonomous local government itself. The California Supreme Court, long a liberal pace-setter, agreed with Serrano that differential property tax bases, coupled with state "equalization" schemes which did not always equalize, could not stand the test of the equal protection clause.[55] In Texas, a three-judge federal district court found equally unacceptable variations in the Bexar County school districts.[56] But "Serrano" and "Rodriguez" dealt with only one side of the distribution of resources: the inter-district differences in school tax bases. Julius Hobson and Andrew Hawkins drew attention to the *intra*-district and the *intra*-municipal distribution of school and nonschool expenditures. In "Hobson" and all its progeny, Judge Skelly Wright ordered the District of Columbia School District to equalize its expenditure patterns *within* the district.[57] The parallel decision affecting municipal non-school services is "Hawkins v. Shaw,"[58] where the fifth circuit found the sharpest possible differences in services from black to white neighborhoods, and held that such discrimination contravened the equal-protection clause.

Had the Burger Court upheld the "Rodriguez" application, the equal-protection clause would have broken through to a whole new category of issues: public taxation, finance, and expenditure, and, no doubt, public service distribution. But like earlier courts in earlier eras, the Burger Court has set about putting the Equal Protection clause to rest again. Despite Archibald Cox's dictum that "once loosed, the idea of equality is not easily cabined," the Nixon Court has determined to prove him wrong by closing the lid on Pandora's fingers.[59] Although some of the Warren thrust in the equal protection direction has been rechanneled into the due process route, the sustained thrust of the equal-protection clause is gone for now. Admittedly, as no specific noneducational service discrimination cases ever reached the Warren Court, we cannot say what might have been. But we can speculate about what will be, and it seems unlikely that the new chief justice and the new court will mandate service equality. The change from Warren to Burger may well be to turn de Tocqueville on his head: What might have been a matter of judicial resolution will now have to become a feat of political power.

Conclusion: A Confluence of Philosophical, Constitutional, and Social Science Issues

If there is a single issue about which philosophical, legal, and social policy perspectives converge, it is the problem of equality. What people *do* get from government, and what they *should* get can be differentially answered. But the issue properly comes to rest where society's crucial resources of security, property, leisure, learning, and residence are concerned, and that is at the urban level. So far, we have drawn freely upon both philosophical and constitutional conceptions of equality. We have even examined locality-specific opinion distributions about who gets what from urban governments. This brief tour has not, it seems clear by now, produced very definitive answers. The problem of equality is as unsettled in the law as it is in philosophy. Nor, for that matter, has the social science evidence we have briefly examined been conclusive. We shall return subsequently to this evidence as we examine patterns of service allocation in our case in point.

In some respects, it is unfortunate that the leading constitutional case on service discrimination has been from Shaw, Mississippi. Whether comparable patterns of service discrimination can be found elsewhere is an empirical question. Shaw, Mississippi, is not Syracuse, San Francisco, or San Antonio. The sheer blatancy of the Shaw precedent resembles the crass discrimination of the Jim Crow era, where public toilets were provided for whites but not for blacks and where high schools were open to whites but not to blacks, and where voting rights were extended to whites but not to blacks. Complex urban systems may produce more complex delivery systems. It is thus more challenging and more important to examine the distribution of public services in elaborated political systems. That is our concern for the next three chapters. In doing so, we focus essentially on two questions: *whether* and *why*. Whether urban services are differentially allocated by social class, race, and political power is the first issue we address. Why one or another pattern of service delivery prevails is equally important. In the next chapter, we develop a detailed plan of analysis for addressing these questions. The answers, it will turn out, are not always as simple as conventional wisdom implies, and will prove to be—like the concept of equality itself—richly complex.

NOTES

1. *The Unwalled City* (New York: Basic Books, 1972), p. 36.
2. "Foreward: Constitutional Adjudication and the Promotion of Human Rights," *Harvard Law Review,* 80 (1966), p. 91.
3. *The Affluent Society* (New York: New American Library, 1958), p. 72.
4. "Egalitarianism: Mechanisms for Redistributing Income," December 8, 1975, p. 86.
5. John Rawls, *A Theory of Justice* (Cambridge, Mass.: Harvard Univ. Press, 1971); Michael Nozick, *Anarchy, States and Utopia* (New York: Basic Books, 1974).
6. *Social Justice and the City* (Baltimore: Johns Hopkins, 1973), p. 52.
7. Cited in Hugo Adam Bedau, "Egalitarianism and the Idea of Equality," in J. Roland Pennock and John W. Chapman, eds., *Equality: Nomos IX* (New York: Atherton, 1967), p. 4.
8. Paul E. Sigmund, "Hierarchy, Equality, and Consent in Medieval Christian Thought," *ibid.,* p. 138.
9. Cited in Sanford Lakoff, *Equality in Political Philosophy* (Cambridge, Mass.: Harvard Univ. Press, 1964), p. 90.
10. *Hawkins v. Shaw,* 437 F.2d 1286 (5th Cir. 1971); Plaintiff's memorandum, *Woo v. Alioto,* November 10, 1970, pp. 8-11.
11. Lakoff, *op. cit.,* p. 6.
12. "Developments in the Law: Equal Protection," *Harvard Law Review,* 82 (1969), p. 1165.
13. 351 U.S. 12, 34 (1956).
14. Richard Herrnstein, *I.Q. in the Meritocracy* (Boston: Little, Brown, 1973), pp. 197-198. For an economist's rebuttal to Herrnstein, see John Conlisk, "Can Equalization of Opportunity Reduce Social Mobility?" *American Economic Review,* 64 (March, 1974), pp. 80-90.
15. James Coleman, et al., *Equality of Educational Opportunity* (Washington, D.C.: Government Printing Office, 1966). For several critiques of the "Coleman report," see Frederick Mosteller and Daniel P. Moynihan, eds., *On Equality of Educational Opportunity* (Cambridge, Mass.: Harvard Univ. Press, 1972).
16. Christopher Jencks, *Inequality* (New York: Basic Books, 1972).
17. *San Antonio Independent School District v. Rodriguez,* 411 U.S. 1 (1973).
18. "The Concept of Equality of Educational Opportunity," *Harvard Educational Review,* 38 (1968), pp. 16-17.
19. "Equal Schools or Equal Students?" *The Public Interest,* (Summer 1966), p. 72.
20. Ira Sharkansky, "Government Expenditures and Public Services in the American States," *American Political Science Review,* 61 (December, 1967), pp. 1066-1077.
21. 468 F.2d 287 (2d Cir. 1972).
22. See the results of the Kansas City experimentation in patrol practices, reported in George L. Kelling, et al., *The Kansas City Preventive Patrol Experiment: A Summary Report* (Washington, D.C.: The Police Foundation, 1974).
23. "Equality of Opportunity—and Beyond," in Pennock and Chapman (eds.), *op. cit.,* p. 233. For a similar argument, see Frank Parkin, *Class, Inequality and Political Order* (New York: Praeger, 1971), pp. 82-88.

24. Lee Rainwater, *What Money Buys: Inequality and the Social Meanings of Income* (New York: Basic Books, 1974), p. 32.

25. U.S. Department of Health, Education and Welfare, ESEA Title I Program Guide No. 44, Guideline 7.1 (March 18, 1968), emphasis added.

26. See Mark Yudof, "The New Deluder Act: A Title I Primer," *Inequality in Education,* 2 (December 5, 1969), pp. 1-2f. So serious were the state and local deviations from the standards, themselves gradually relaxed over the years, that in 1972, the United States Office of Education instituted legal actions to recover monies from several states and districts which were in violation of the supplement-not-supplant rule.

27. Rainwater, *op. cit.,* p. 168.

28. *Fantasy and Science Fiction,* (October, 1961), p. 5.

29. See, e.g., Robert Fogelson, "From Resentment to Confrontation: The Police, the Negroes, and the Outbreak of the 1960s Riots," *Political Science Quarterly,* 83 (June, 1968), pp. 217-247; Joel D. Aberbach and Jack L. Walker, "The Attitudes of Whites and Blacks toward City Services: Implications for Public Policy," in John P. Crecine, ed., *Financing the Metropolis* (Beverly Hills: Sage, 1970), pp. 519-537; and the *Report* of the National Advisory Commission on Civil Disorders.

30. Howard Schuman and Barry Gruenberg, "Dissatisfaction with City Services: Is Race an Important Factor?" in Harlan Hahn, ed., *People and Politics in Urban Society* (Beverly Hills: Sage, 1972), pp. 396-392.

31. For a more detailed report on the urban observatory data, see Floyd J. Fowler, Jr., *Citizen Attitudes Toward Local Government, Services and Taxes* (Cambridge, Mass.: Ballinger, 1974).

32. Schuman and Gruenberg, *op. cit.*

33. (New York: Viking, 1961).

34. John D. Owen, "The Distribution of Educational Resources in Large American Cities," *Journal of Human Resources,* 7 (1972), pp. 26-38; Richard A. Berk and Alice Hartman, "Race and District Differences in Per Pupil Staffing Expenditures in Chicago Elementary Schools, 1970-1971," Center for Urban Affairs, Northwestern University, (June, 1971). See also the data reported in *Hobson v. Hansen,* 269 F. Supp. 401 (D.D.C. 1967), aff'd sub nom. *Smuck v. Hansen,* 488 F.2d 175 (D.C. Cir. 1969).

35. See e.g., Katzman, *op. cit.:* David Lyon, "Capital Spending and the Neighborhoods of Philadelphia," *Business Review* (of the Federal Reserve Bank of Philadelphia), (May, 1970), pp. 16-27; and Frank S. Levy, Arnold J. Meltsner and Aaron Wildavsky, *Urban Outcomes: Schools, Streets and Libraries* (Berkeley: Univ. of California Press, 1974).

36. Allen S. Mandel, "The Allocation of Resources inside Urban and Suburban School Districts: Theory and Evidence," unpublished Ph.D. dissertation, Department of Economics, University of Michigan, 1974.

37. *Hadnott v. City of Prattville,* 309 F. Supp. 967, 972 (1970).

38. Lowell A. Martin, *Library Response to Urban Change: A Study of the Chicago Public Library* (Chicago: American Library Association, 1969), p. 40.

39. *Hobson v. Hansen,* 269 F. Supp. 401, 517 (1969). We may note in passing that Appeals Court Judge Danaher, dissenting when his brethren reaffirmed Wright's decision, dutifully quotes part of the above passage, omitting, however, the last two sentences. His dissent was joined by his colleague Warren Burger.

40. The number of law journal articles is always a good index. The seminal one is Gershon M. Ratner, "Inter-neighborhood Denials of Equal Protection in the Provision of Municipal Services," *Harvard Civil Rights-Civil Liberties Law Review,* 4 (Fall, 1968), pp. 1-64. See also Note, "Equal Protection Across the Tracks—*Hawkins v. Town of Shaw,*" *University of Pittsburgh Law Review,* 32 (Summer, 1971), pp. 555-579; Note, "The

Right to Adequate Municipal Services: Thoughts and Proposals," *New York University Law Review*, 44 (1969), pp. 753-774; Comment, *"Hawkins v. Town of Shaw*—Equal Protection and Municipal Services: A Small Leap for Minorities but a Giant Leap for the Commentators," *Utah Law Review*, (Fall, 1971), pp. 397-404; Daniel W. Fessler and Lucy S. Forrester, "The Case for the Immediate Environment," *The Clearinghouse Review*, 4 (May and June, 1970); Robert L. Graham and John H. Kravitt, "The Evolution of Equal Protection—Education, Municipal Services, and Wealth," *Harvard Civil Rights-Civil Liberties Law Review*, 7 (June, 1972), pp. 103-213; Note, "Equalization of Municipal Services: The Economics of *Serrano* and *Shaw*," *Yale Law Journal*, 82 (November, 1972), pp. 89-122; and Robert L. Lineberry, "Mandating Urban Equality: The Distribution of Municipal Public Services," *Texas Law Review*, 53 (December, 1974), pp. 26-59.

41. *Bair v. Mayor and City Council of Westminster*, 243 Md. 495, 221 A.2d 643, 645 (1966).

42. *Veach v. City of Phoenix*, 102 Ariz. 195, 427 P.2d 335, 337 (1967).

43. *Travaini v. Maricopa Co.*, 9 Ariz. App. 228, 450 P.2d 1021, 1022 (1969).

44. *Mo. Stat. Ann.* 71.015 (1959). For an example of the elaborate analysis of one court assessing the quality of municipal services, see *City of Gape Girardeau v. Armstrong*, 417 S. W.2d 661 (Mo. Ct. App. 1967).

45. The standard source on the concept and development of the equal protection clause is Joseph Tussman and Jacobus tenBrock, "The Equal Protection of the Laws," *California Law Review*, 37 (September, 1949), pp. 341-381. For a review of the Warren Court and the clause, see "Developments in the Law: Equal Protection," *Harvard Law Review*, *op. cit.*

46. "Invidious Discrimination: Justice Douglas and the Return of the 'Natural-Law-Due Process' Formula," *UCLA Law Review*, 16 (1969), p. 716.

47. Tussman and tenBrock, *op. cit.*, p. 345.

48. *Ibid.*, p. 354.

49. *McLaughlin v. Florida*, 379 U.S. 184, 191-192 and 196 (1964).

50. *Reynolds v. Sims*, 377 U.S. 533 (1964).

51. *McDonald v. Board of Election Commissioners of Chicago*, 394 U.S. 802, 807 (1969), emphasis added. Wealth was also singled out as a suspect classification in *Roberts v. Lavallee*, 389 U.S. 40 (1957); *Griffin v. Illinois*, 351 U.S. 12 (1966); *Douglas v. California*, 372 U.S. 353 (1964); *Serrano v. Priest*, 96 Cal. Rptr. 601 (Sup. Ct. Cal. 1971); and *Hobson v. Hansen*, 269 F. Supp. 401 (D.D.C. 1967).

52. *Serrano v. Priest*, 96 Cal. Rptr. 601, 613 (Sup. Ct. Cal. 1971).

53. See the famous *Brown* decision on education. On parks, see, e.g., *Hadnott v. City of Prattville*, 309 F. Supp. 967 (M.D. Ala. 1970); *Hague v. CIO*, 307 U.S. 496 (1939); *Watson v. Memphis*, 373 U.S. 526 (1963).

54. *Korematsu v. U.S.*, 323 U.S. 214 (1944).

55. *Serrano v. Priest*, 96 Cal. Rptr. 601 (Sup. Ct. Cal. 1971).

56. *Rodriguez v. San Antonio Independent School District*, 330 F. Supp. 280 (1971); reversed by the Supreme Court on March 21, 1973, 411 U.S. 1 (1973).

57. *Hobson v. Hansen*, 269 F. Supp. 401 (D.D.C. 1967), aff'd sub nom., *Smuck v. Hansen*, 408 F.2d 175 (D.C. Cir. 1969).

58. 437 F.2d 1286 (1971).

59. For an analysis of the changing decisions on equal protection, see Wallace Mendelson, "From Warren to Burger: The Rise and Decline of Substantive Equal Protection," *American Political Science Review*, 66 (December, 1972), pp. 1226-1233.

Chapter 3

HOW TO MEASURE SERVICE DISTRIBUTION

One of America's four unique cities.

> *Will Rogers*

Them that has, gets.

> *Old Cliché*

Our principal hypothesis is that not all urbanites share alike in the largesse of government policy outputs. Our test case is San Antonio, Texas. While there are, as we shall see, several reasons to make it an attractive example, it was chosen as mountains are climbed (and as most research sites are picked), simply because it was there. Unless its pattern of urban service distribution proves utterly random, it should be possible to identify the correlates, if not the causes, of inequalities if and where they exist. Such an identification should take us some ways toward suggesting elaborations in other locales. In this chapter, we outline four principal hypotheses which may singly or jointly explain service distribution patterns. We characterize these as the racial preference, the class preference, the power elite, and the ecological hypotheses. Also introduced, but not elaborated until later, is a fifth hypothesis, which we shall call the bureaucratic decision-rule hypothesis.

The Case

San Antonio is not one of those cities overrun by social scientists. Other places—obviously New York, Chicago, and Boston, but also lesser spots like Muncie, Durham, New Haven, and Oakland—have captured the imagination of social scientists, or have simply been more convenient to study. San Antonio, however, is the nation's eleventh city in size,[1] it is interesting in its own right, and historic far beyond the Alamo. Electorally, an elite which would make Floyd Hunter's Atlanta seem like fraternity politics, long dominated the city. Ethnically, it contains few blacks, but more than half its population is Chicano. Economically, the city is heavily dependent upon its military bases (Fort Sam Houston, Kelly and Lackland Air Force Bases), but is fundamentally a poor one. Among the nation's largest cities, San Antonio brings up the rear on nearly every measure of economic well-being. San Antonio is hardly typical, but it is thoroughly interesting.

For illustration, the data in Table 3.1 offer a handy comparison of San Antonio with other larger (over 200,000 in population) Standard Metropolitan Statistical Areas. The fact of the matter is

Table 3.1: Statistical San Antonio*

		San Antonio	200 Largest SMSAs
I.	*Population*		
	a. Total SMSA	864,000	N.A.
	b. Central City	654,000	N.A.
	c. Percentage CC is of SMSA	75.7%	45.5%
	d. Growth rate, 1960-70	11.3%	N.A.
	e. Population per square mile	4414	5976
II.	*Ethnicity*		
	a. Percentage nonwhite in CC	8.6%	23.9%
	b. Percentage Mexican stock in CC	52.1%	N.A.
III.	*Income*		
	a. Percentage below census poverty line	16.0%	10.7%
	b. Median family income	$7981	$9590
IV.	*Housing*		
	a. CC owner occupancy	62.4%	46.8%
	b. Median value of owner occupied housing	$11,600	$16,500
	c. Percentage housing units with more than 1.01 persons per room	16.3%	8.0%

*The soure of most of the data above is U.S. Bureau of the Census, *Metropolitan Area Statistics* (Washington, D.C.: Government Printing Office, 1972).

that San Antonio ranks near the extremes of nearly every indicium of metropolitan demography. Specifically,

1. San Antonio is high (ranking seventh) in the proportion of people in the SMSA residing in the Central City (CC).

2. San Antonio is extremely poor. Only nine of the 157 SMSAs with populations over 200,000 have lower median family incomes than San Antonio.

3. Although San Antonio is low in the proportion of its black population, it is extremely high in the proportion of its population of Mexican stock. In fact, in absolute numbers, only New York and Los Angeles exceed it in numbers of persons of Spanish heritage.

4. The city's rank in home ownership is quite high, although the homes are of relatively low value. Overall, San Antonio's housing conditions are poor indeed. Only two large SMSA's (El Paso and Miami) score worse on the conventional measure of overcrowding, i.e., more than 1.01 persons per room.

We can hardly contend that our case in point is a typical American metropolitan area. Yet in some ways, it is the distinctiveness of the case which makes it appealing. If poor persons in the city are discriminated against by the public sector, there are plenty of poor in San Antonio to discriminate against. Poverty there is tri-ethnic, including Anglos, blacks, and Chicanos, although the latter are numerically dominant.[2] Fortunately, too, for our purposes, the minimal suburbanization of the city permits a considerable range of variation in class and ethnic variation within the unit of analysis. This SMSA does not present the typical—or stereotypical—case of the "huddled masses" in the central city and the Gatsby-like "gilded ghettos" of the suburbs.

Politically, the city is also sui generis. We may begin with its "power structure." Granted that no one has yet developed a failsafe method of measuring the concentration of power in American cities, few would claim that San Antonio follows the pluralistic model of Dahl's New Haven. The most hard-headed pluralist could not ignore the potent electoral influence of the Good Government League (GGL). The electoral success of the organization would rival that of the strongest old-style urban machine. Since its inception in 1954 and until the municipal elections in

1973, the GGL was victorious in seventy-eight of eighty-one local council races.[3] Only three of forty elected council members during this period were elected without its endorsement. One of these (Councilman, now Congressman Henry Gonzales) was elected without GGL support, but without their active opposition. The other two defeated GGL-slated candidates, but were themselves ousted at the next election. If pluralists and elitists would agree on anything, surely they would agree that San Antonio's Good Government League packs a considerable electoral punch. One may even say that the GGL is a power structure's power structure.

Organizationally, the GGL is not a casual, "crowd at the civic club," political clique. It is a formal, permanent organization, with stable leadership, solid financing, and an elected board of directors.[4] It is hardly a secret cabal. In fact, it maintains a suite in an office building partly owned by long-time GGL Mayor W. W. McAllister. It dutifully reports its finances to the city clerk. For instance, in 1971, it reported outlays of $152,434.64, three-quarters of which went to advertising for its slate. The figure represented almost three times the combined expenditures of all other candidates. The most any opposing candidate mustered was $11,032.26, far short of the average of $17,000 per GGL candidate. The GGL is basically an electoral organization, a sort of upper-class political machine, officing not in Tammany Hall, but in a savings and loan association, whose electoral wonders are impressive to behold. In a head-on battle with it, even Mayor Daley might have blinked.

The candidates supported by the GGL, like their board of directors, represent a rather special segment of the community's population. Most have historically been drawn from high status occupations. Consistent with the axiom that "many are called but few are chosen," numerous neighborhoods have had candidates running for local office, but San Antonio's at large system has favored certain neighborhoods over others. Between 1955 and 1971, fully eighty-eight of the city's one hundred and thirteen census tracts had no councilmanic representation. Indeed, chances were that if a neighborhood had one councilman, it had several. One tract (the one with the highest median family income in the city) held eight of the thirty-six seats during the period

1965-1971. "Represented tracts" scored higher on every available measure of socioeconomic status. Ethnically, councilmen have tended to be disproportionately Anglo. Of the thirty-six seats open during the late 1960s, only two were held by blacks, and only five were held by persons with Spanish surnames. San Antonio thus follows a pattern of electoral dominance by white, upper-middle class groups, long familiar in many at large, nonpartisan cities.[5] The at large, nonpartisan system, though, is not the sole explanation. The GGL itself, its sophisticated leadership, organization, and financial base, contributes to making San Antonio one of the nation's "unique cities," not only for tourism, but for political scientists as well. Our concern is not, though, for the politics and power structure of a large city. Rather, we focus on its public policies. To us, San Antonio is an instant case, wherein we may examine the burdens and benefits of urban policy. In the rest of this chapter, then, we develop more systematically the hypotheses of the study and outline the research itself.

Five Contending Explanations

If urban public services are differentially allocated to various neighborhoods, any number of explanations may be found. Some allocative choices may be unintentional, accidental, or beyond the ready control of decision-makers. Other allocations may reflect the natural lag of public sector expansion catching up with private sector growth, when, for instance, a new subdivision is built but public services have not yet fully expanded to accommodate it. Still other allocations may take on a malevolent coloration, as, for example, policemen being slow to respond to calls from a minority neighborhood.

Much of the conventional wisdom about urban services holds that service distribution is a function of the discrimination against the urban "underclass." According to this hypothesis, one or another advantaged group suffers from a series of ills arising from the conscious or unconscious decisions of public officials. However service quality and quantity is measured, such groups come up on the short end of the service stick. From this conventional wisdom, we shall draw three hypotheses, the race preference, the class preference, and the power elite explanations. These explanations are really quite similar in that they each identify some group (minorities in general, the poor, or the powerless) which fails to share in the largesse of urban government. There are, however, two additional hypotheses of a different sort. The ecological hypothesis suggests that service distribution is a function, not of overt or even covert discrimination, but of the ecological attributes—age, density, and such—of particular neighborhoods. The bureaucratic decision-rule hypothesis, finally, suggests that all allocations are results of convenient, time-saving decisional premises of bureaucracies, whose allocative consequences are little more than accidental by-products of internal decision-making.

THE RACE PREFERENCE HYPOTHESIS

All three "underclass" hypotheses share common features. They assume that, if anything, public services are probably more important to the powerless, the poor, or the minorities than to more advantaged groups. The reasons are not obscure. Theodore Lowi has observed that every public service has its private counterpart, including private schools, private security services, private parcel services, private garbage collectors, and so on.[6] But both economic and locational advantages may enable affluent citizens to avail themselves more readily of private sector services. Public pools are less crucial if one's family or friends have a private one; public parks are more important to those without spacious lawns; public libraries are more essential to homes where books are a luxury; fire protection is more essential to those with old frame than new brick houses; even in education, the advantaged child may find that family resources will sustain him or her in a poor school, while less advantaged children need good schooling to overcome initial disabilities. In other words, what the public sector does not adequately provide, the advantaged household can typically secure by other means; the disadvantaged family will have to do without.

All this is commonplace argument. What has never been clear enough, though, is that service delivery patterns contribute not only to immediate life satisfactions and need fulfillment, but also affect the pattern of race and class segregation in the metropolis. Housing segregation is more than the unhappy confluence of housing and job discrimination. In a famous article, Tiebout contends that citizens choose a residential location within the fragmented governments of a metropolis partly on the basis of their preferred mix of taxes and services provided by different municipalities.[7] This logic also applies to intra-municipal locational divisions. Yet residential choices for racial and income minorities have not been entirely voluntary. Instead, urban governments can and have used service location and quality to foster segregated housing patterns. Especially when segregation was the dominant moral and legal mode of race relations, *it became essential for minority groups to cluster together if they were to receive any services at all.* With white services denied them, nonwhites who expected to enjoy schools, parks, sewers, and other services, had to flock together in segregated areas.

The case of Austin, Texas, is instructive if not typical.[8] As the city developed around the turn of the century, it contained three major pockets of black settlement (Wheatsville, Clarksville, and "East Austin") located in different quadrants of the city. The early years of the twentieth century marked the beginnings of publicly constructed streets, sewers, water mains, parks, and lighting systems. (It is, incidentally, too easy to assume that public services have long been with us. Even in Boston, neither common schools nor regular police protection began until the 1830s.) By 1927, Austin felt compelled to hire a private consulting firm, Koch and Fowler, to develop plans for the orderly expansion of public facilities and utilities. The issue of racial segregation was not overlooked by the consultants, who concluded that

> there has been considerable talk in Austin, as well as other cities, in regard to the race segregation problem. *This problem cannot be solved legally under any zoning law known to us at present.* Practically all attempts of such have been proven unconstitutional.
>
> In our studies in Austin we have found that the negroes [sic] are present in small numbers in practically all sections of the city, excepting the area just east of East Avenue and south of the city cemetary [i.e., "East Austin"]. This area seems to be all negro population. *It is our recommendation that the nearest approach to the solution of the race segregation problem will be the recommendation of this district as a negro district; and that all the facilities and conveniences be provided to the negroes in this district, as an incentive to draw the negro population to this area.* This will eliminate the necessity of duplication of white and black schools, white and black parks, and other duplicate facilities in this area.[9]

Or, in other words, if a black family intended to use the public parks, enjoy streets in its neighborhood, hook up to a public sewer, and attend school close to home, it would be well-advised to move the area of "all negro population." This is the "magnet" aspect of urban service distribution. Attracting minorities to some neighborhoods and repelling them from others where they could not secure services, public service distribution made an independent contribution to residential segregation. Indeed, shortly after adoption of the Koch and Fowler plan, the city parks board consulted the new plan and purchased its first "colored playground" deep in the heart of East Austin. No other area of the city was seriously considered. Whether such practices were commonplace in the development of urban services in every city, we cannot say, largely because historians as well as political scientists have neglected the study of urban public services.

The instances of suspected or demonstrated discrimination in the distribution of urban services on racial bases is a story too long to be fully recounted here. It does, however, warrant the first hypothesis, the race preference hypothesis. San Antonio will be an acid test. It contains forty-three of one hundred and thirteen census tracts which are predominantly—more than 80 percent—minority, although the principal ethnic group is Chicano rather than black.

THE CLASS PREFERENCE HYPOTHESIS

The class preference hypothesis takes a more inclusive posture than the race preference hypothesis, holding that the economically disadvantaged in general—Anglo, Chicano, and black—are shortchanged. In her study of disparities in school services in Detroit, for example, Patricia Cayo Sexton observed that "the problems we describe here . . . are principally social class problems, not racial problems." [10] The question of whether class or race is a more important explanation of service discrimination is not one of merely idle academic curiosity. There is a world of constitutional difference between discrimination as a function of income alone and discrimination as a function of race. The former has a tenuous hold on unconstitutionality, while the latter patently affronts the fourteenth amendment. [11]

THE POWER ELITE HYPOTHESIS

It is hardly unheard of—Watergate and similar sins are merely the latest reminders—for those in power to favor themselves and their friends. The most common strain in the "power elite" theories of urban politics is that elites rule in their own interest, advantaging themselves and their fellows with contracts, tax breaks, appointments, service preferences, and other public sector goodies. Elites themselves are not scattered randomly about the sociospatial landscape. As Floyd Hunter observed of Atlanta, "there tends to be a clustering of residential quarters of the leaders. . . . They meet in common places and live in close proximity to one another. This is structurally significant." [12] As in Atlanta, so, too, in San Antonio. Members of the community political elite—provisionally defined herein as members of the GGL board of directors, members of the city council, and department heads within city government—do cluster in particular census tracts. And it does not require a very radical interpretation of political power to hypothesize that those neighborhoods heavily populated by the influentials may have more clout in securing urban services. Kasperson, for example, find some evidence that neighborhoods which most vigorously support Chicago's machine, get the best public services Chicago provides.[13] That is, in fact, our third hypothesis about the distribution of urban services.

THREE OF A KIND: THE UNDERCLASS HYPOTHESIS

The race preference, class preference, and power elite hypotheses are, in a sense, three of a kind, each representing an "underclass" explanation of service delivery. In every case, a zero-sum assumption pervades the argument and each assumes a distributive principle consisten with the old maxim that "them that has, gets." Each assumes also a dominant causal variable (race, class, or power) in determining policy outputs. Unfortunately, it is very difficult empirically in San Antonio (or elsewhere) to decide which, if any, of these underclass explanations best describes urban service distributions. The inter-correlations among race, class, and power are sufficiently powerful to complicate any effort to measure their independent effects. Table 3.2 shows the inter-

Table 3.2: Correlations among Ethnicity, Poverty, and Political Power,
San Antonio Census Tracts, 1970*

Indicators	Percentage Minority	Percentage Poverty Families	Number of Political Elites
Percentage minority	1.00	−.84	−.45
Percentage poverty families	−	1.00	−.39
Number of political elites	−	−	1.00

*For a precise definition of these indicators, see Table 3.3 below and the textual
description associated with it.

correlations among the three indices of poverty, ethnicity, and
political power. The size of the relationships, especially that
between ethnicity and poverty, are significant enough to caution
against disentangling the effects of race, class, and power. To
secure a real test, one would need nine classes of cases, e.g.,
neighborhoods which were rich, minority, and powerless; poor,
minority, and powerless; rich, nonminority, and powerless; etc. No
American city—not even New Haven, Dahl's bastion of pluralism—
offers such a range. We will do what we can to disaggregate the
conjoint effects of these inter-correlations, but we cannot reach
any definitive conclusions about which of these three of a kind
hypotheses works best to describe distributional patterns—
assuming, of course, that the underclass theory works at all.

THE ECOLOGICAL HYPOTHESIS

The underclass hypotheses assume a Machiavellian malevolence
to urban decision-making, at least from the point of view of the
underclass itself. Inequalities in urban services are explained in
terms of bad motives, bad values, or even bad people. There are,
however, less conspiratorial explanations of urban decision-
making, and they form the basis of our last two hypotheses, the
ecological and decision-rule explanations. The first is directly
testable with the quantitative data available, the second, unfortunately, can only be indirectly confronted.

The ecological hypothesis comes in different guises, but its
underlying assumption is that attributes of neighborhoods other
than their prima facie political ones—race, class, and power—
account for distributional decisions. In particular, the ecological

traits of a neighborhood—*ecological* being defined here broadly enough to encompass geographical aspects, age and the density of an area, and a number of other traits—determine service delivery. Some ecological traits are purely fortuitous accidents of geography or historical accident. The physical properties of a land parcel may make it ideal for public parks, or even for dumps and sewerage treatment plants. Historical accident may foreclose later options. In San Antonio, for example, the city's major park, Brackenridge, a magnificent miniversion of Mexico's Chapultapec, happens to be located nearer the richer than the poorer part of town. There is nothing manipulative, mysterious, or malevolent about this, however. The reason is a historical accident, involving the donation in the 1880s of a large tract of land by a wealthy benefactor. At the time, it was distant enough from everyone not to advantage any particular income groups. Only a short hop up Interstate 35, in Austin, the opposite accident occurred, where the major park is closer to the poor than the rich. But neither benefactor could possibly have been prescient enough to contemplate purposeful discrimination.

A number of ecological factors are not accidental, but reflect conscious, defensible determinations. Density and age of neighborhoods are factors which might well figure in distributional patterns. Quite possible, a curvilinear relationship best describes the correlation between age of neighborhood and services received. We might expect, in parks and libraries, for example, the worst facilities in the oldest neighborhoods. Older areas were often built before planners and public officials paid much attention to the niceties of open space. Almost certainly, parks there do not reflect up-to-the-minute standards for recreational design and facilities. Add to that the problem of density and the high cost of land in old, core areas, and the issue becomes even more complicated. At the other extreme, of course, is the newest area of the city. These "developments" or "subdivisions" are usually on the periphery of the city and there is naturally a certain lag between development of a neighborhood and the construction of public facilities to serve it. Indeed, it would be a foolish use of public resources if, every time a few new homes are put up, expensive parks, pools, fire stations, libraries, and freeways, were put up to service them. It is not, therefore, an unreasonable supposition that ecological attri-

butes of neighborhoods may also account for the distribution of
public services.

THE DECISION-RULE HYPOTHESIS

The urban bureaucracy and its decisional processes remain the
terra incognita of urban political analysis. The dominant community power studies of the 1950s and 1960s ignored the public
bureaucracy almost completely. As Francis Rourke remarked of
Dahl's *Who Governs?*, one might never guess from reading the
book that New Haven had a public bureaucracy.[14] Since the spate
of power structure studies, fortunately, some attention has been
given to the patterned routines of bureaucrats as well as to the vast
discretion which they exercise. The regularities of bureaucratic
behavior are part of what Sharkansky calls the "routines of
politics."[15]

The quantitative studies of Crecine at the urban level, of Sharkansky at the state level, and of Wildavsky and others at the federal
level show the high degree of predictability of budgetary decisions.[16] Not only is there a powerful correlation in expenditure
patterns from one year to the next—the phenomenon called
"incrementalism"—but there are certain simple rules which collectively can be used to predict expenditure patterns of governments.
Crecine, for example, built a simulation model encompassing a set
of rules for various actors on the urban budgeting system, and
predicted with astonishing accuracy the levels of actual expenditure patterns in Detroit, Cleveland, and Pittsburgh. Crecine
supports a model of "internal" influences on decision-making,
where budgetary policies are almost immune from outside forces
like group pressure, citizen attitudes, and political considerations.
The implication of such analyses is that public bureaucracies
behave in highly structured, routinized, and deterministic fashions
in making public choices. Collectively, these studies of public
budgeting point up a set of decision-rules which make for a rather
rigid, closed system.

Recently, however, there has been a resurgence of attention to
administrative discretion. Superficially at least, studies of this
genre appear plainly to contradict the image of the bureaucracy
derived from the budgetary research. In the administrative discretion literature, what is clearest is the wide range of choice for both

top administrators and the lowest functionaries. John Gardiner and Martha Derthick have each shown what vast power is concentrated in the hands of leaders of the police department and the welfare bureaucracies in determining which rules are enforced.[17] Gardiner's study of inter-urban variations in traffic ticketing policy showed that Dallas policemen in a single year wrote twenty-four times as many tickets as did Boston policemen, even though the two cities were of comparable size. The most important variable explaining such differences was the attitude of the police chief toward the importance of traffic law enforcement. Change the police chief to one with different beliefs about the traffic control function, and the number of tickets rises or falls dramatically.

It is not only at the top of the bureaucratic ladder that discretion is apparent. If the policeman in Michigan is required to enforce an estimated 50,000 laws, he will obviously need to exercise some discretion. Gilbert and Sullivan held that "the policeman's lot is not a happy one," and one reason is because the vast discretion of the policeman is exercised in the most abrasive of environments. James Q. Wilson writes that the maintenance of order involves "sub-professionals, working alone, [exercizing] wide discretion in matters of utmost importance (life and death, honor and dishonor) in an environment that is apprehensive and hostile."[18] Similarly, welfare workers find plenty of room for judgment despite the voluminous rule books intended to routinize their responsibilities. Collectively, Lipsky describes these bureaucratic agents—the policemen, the welfare workers, the parole officers, the lower court judges, and so forth—as the "street level bureaucrats," whose great power is coupled with considerable independence about the operational application of public policy.[19]

There appears to be, however, a tension between these two images of the bureaucracy. In one, the bureaucracy is portrayed as rigid, highly structured by its decision-rules, rather closed to external forces, and relatively predictable. In the other, the bureaucracy is soft, enjoys considerable discretion, is idiosyncratic, and individual bureaucratic decisions are unpredictable. Such inconsistency may be more apparent than real. Budgetary decisions probably do stand at one extreme. For reasons related to the quantitative and easily divisible character of money, fiscal decisions are easily routinized. At the other extreme stands the chance encounter between the beat patrolman and the gang of teens in the street. Even then, the case is probably not so idiosyncratic that no routine is involved in its handling. (A thoroughly idiosyncratic case was the San Antonio police officer who dutifully filled out an action report on the case of a lion in the back of a truck parked at a motel.) Bureaucracies do enjoy considerable discretion, but their discretion is made more manageable by the development of decision-rules and routines. While not immutable, such rules are economizing devices. The rules by which garbage is collected, police patrols are assigned, street lights are installed, playground equipment distributed, fire stations are built, and new books allocated to branch libraries are all decisional routines affecting the distribution of public services. Their inception may be entirely internal in character. They may affect the distribution of public services even though they are mere by-products of bureaucratic convenience. Their latent functions, however, may be more important than their manifest functions. And, while we cannot readily quantify decision-rules of public bureaucracies as we can the racial, economic, political, and ecological traits of urban neighborhoods, we can at least give them due attention.

THE HYPOTHESES: AN OVERVIEW

We operate with five working hypotheses, the first three of which can be subsumed under the umbrella of an "underclass" explanation. Briefly stated, they are

1. that the quantity and/or quality of urban services are positively related to the proportion of Anglos in a neighborhood population (the race preference hypothesis);

2. that the quantity and/or quality of urban services are positively related to the proportion of the neighborhood population which is of higher socioeconomic status (the class preference hypothesis);

3. that the quantity and/or quality of urban services are positively related to the proportion of the neighborhood population occupying positions of power in urban government (the power elite hypothesis);

4. that the quantity and/or quality of urban services are functions of ecological aspects of urban neighborhoods, including but not limited to their age, density, geographical character, and residential-commercial mix (the ecological hypothesis); and

5. that the quantity and/or quality of urban services are primarily functions of bureaucratic decision-rules made to simplify complex allocations of administrative time and resources (the decision-rule hypothesis).

Obviously, there is, both statistically and theoretically, much "multicollinearity" in these various hypotheses. We cannot easily isolate the effects of race, class, and power, and these, in turn, are not unrelated to ecological traits of urban areas. The effort, however, is certainly worthwhile. It makes a considerable difference, both legally and politically, if service discrimination—assuming it exists at all—is keyed to race or to class, or if it is a function of neither. Service distribution has so far been relatively resistent to measurement. Some of the reasons why will become apparent in the next section.

Some Problems in Distributional Analysis

There are several good reasons why we know more about what governments spend than what they do for (and to) their citizenry. Not the least of these is the ease of data gathering. The census bureau is a ready-made research repository, but the interests of economists and, to a lesser extent, demographers, have always received better representation there than the interests of political scientists. But data shortages are not the only—perhaps not even the most important—limitations to distribution analysis. We can identify at least seven such problems:

1. The records problem;

2. The problem of measurement;

3. The fallacy of partial analysis;

4. The ecological fallacy;

5. The problem of public goods; and

6. The problem of externalities.

RECORDS

The availability of data on the distribution of urban services is usually a function of bureaucratic proclivities for record keeping. But, as the economist Werner Z. Hirsch has observed, "relatively little is known about the actual distribution of urban public services by location, race, religion, income class, or other categories—mainly because records are not kept in these terms."[20] Carl Shoup adds that "the laws providing for the service are silent in this respect; the authorizing or appropriating committees of legislatures do not discuss it; budgets submitted by the executive say nothing about how a given service is to be distributed among the users." [21] Sometimes, it may be advantageous to a bureaucracy *not* to retain careful records. A streets department which maintains no records of maintenance cannot so readily be proved guilty of bias. A school board which maintains few records about individual school quality cannot so easily be attacked for racial favoritism. More typically, however, shoddy record keeping is in the nature of bureaucracies. Some agencies are zealous pack rats about data; others collect practically nothing. In San Antonio, the police department maintains a sophisticated data system, yet the public works department does not know how many unpaved streets are found in the city. The fire department accounts meticulously for every hook, ladder and hose, but fails to record how long it requires a company to arrive at the scene of a fire.

These problems, however, can easily be overdrawn. The records are rarely ready-made for the researcher, but chances are that no public service is utterly immune to distributional research because of lack of data. If the data do not exist directly, the researcher can often secure them by direct observation. For example, the Urban Institute once used "roughometers" and blindfolded judges' perceptions of ride comfort to evaluate conditions of streets. [22] A bigger obstacle to distributional research than lack of records is reluctance by the researcher to use such records as do exist.

MEASUREMENT

A magnificent data base would not necessarily solve the problem of measurement. Some of the impact of the policy analyst is blunted by internecine battles over measurement, at times so serious that the policy researcher seems to resemble the professional expert witness who can be found on either side of a case. Nowhere are the potential pitfalls of measurement problems more apparent than in the debate surrounding the Coleman report. [23] If in a rich and well-researched field like education, measurement problems can be so acute, we may assume that they will be trebly serious in an incipient endeavor like urban service measurement. Distributional analysis often grapples with imperfect data, typically generated for internal purposes and aggregated into measures often imperfectly reflecting concepts purportedly being measured. Such measurements are typically chipped from heavy stone, and rarely molded from fine clay.

In its most general sense, the "distribution" of public services includes not only attributes of the services, facilities, and personnel, but also the consumption of services by the population. Yet, admitting consumption variables into the distribution equation will quickly involve the researcher in problems of motivation, information, desires, and interests of population groups. This knowledge invariably requires more delicate instruments than most researchers have available. Moreover, the motivations and interests of citizens are relatively further outside the control of policy-makers than the attributes of service facilities and personnel. A municipal government can be readily faulted for providing inadequate facilities to some of its citizens; if the facilities, however, are adequate, but underutilized, it is much harder to hold officials derelict in their responsibilities. We emphasize in this study the characteristics of the services themselves rather than the consumption habits of citizens, although we shall, at times, relate indicia of quality to service utilization.

Almost all measures of services fall into the general categories like location, promptness, personnel, quality, and delivery. Each of these dimensions is shown in Table 3.3 together with typical indicia. [24] Not all dimensions will be relevant to each particular service. It matters little, for example, where a police station is located (dimension 1) if speed or response time (dimension 6) is high; so long as the fire department is prompt, one does not pay much attention to the demeanor of firemen. Nor will indicia necessarily be available even where they are relevant. Taking the list as suggestive, it should be possible to make some estimates about the overall pattern of service distribution. Shortly we will show how the problem is operationalized in this particular analysis.

THE FALLACY OF PARTIAL ANALYSIS

For any social function (law and order, learning, fire protection, recreation, etc.), there are myriads of private and public elements contributing to both cost and performance. Take recreation. The sum total of the recreation function is obviously assumed pri-

Table 3.3: Dimensions of Policy Distributions with Indicia

	Dimension	Indicia
1.	Location of public facilities	Distance of neighborhood points to desired and/or undesired facilities; access via transportation to facilities.
2.	Crowding of facilities	Ratio of usage to population served; water pressure at peak periods.
3.	Quality of facilities	Structural condition; aesthetic condition; deterioration.
4.	Expenditure per service	Pupil expenditure; library expenditure as a ratio of population served.
5.	Quality of service personnel	Training; ability; experience.
6.	Promptness of service	Police response time; fire department response time; speed of welfare eligibility decisions; snow removal.
7.	Regularity of service	Frequency of garbage collection; street cleaning; police patrol.
8.	Manpower/client ratio	Ratio of professional staff to client (e.g., librarians to population served; pupil teacher ratio).
9.	Eligibility limitations	E.g., possession of library card; limits on welfare, food stamp recipients.
10.	Demeanor of service personnel	Behavior of personnel; conduct with clients; accessibility of service employees.

marily by the private sector (presumably, in American households, by the television set). However large the recreation service investments by local, state and national governments, they are trivial when compared to the contributions of the private sector. If, therefore, we think of government as producing not merely services, but as contributing to the satisfaction of some goal-states, then its performance is conditioned by numerous elements beyond its control. To examine only one element of a production function, or only one fraction of a total cost, is to commit the fallacy of partial analysis.

To put the matter another way, the activities of a given service agency seldom, if ever, solely determine the degree to which the environmental state which is the agency's objective (a secure community, an educated citizenry, etc.) prevails. With respect to police departments and their manifest goal of a secure community, Elinor Ostrom notes that while the activities of a police department "contribute to the security of a community, *it is almost never the sole contributor to this state of affairs.* The degree of security enjoyed by a community is created by individuals interacting with one another within a set of institutional arrangements. . . . Included among these are employment markets, product markets, housing markets, welfare programs, educational systems, community organizations, court systems, penal systems, and the police."[25] An even simpler case is the fire protection function. The total function of fire protection is produced, inter alia, by (1) the facilities and personnel of the fire department; (2) the quality of city building inspection ordinances and code enforcement; (3) the quality and location of public thoroughfares along which departmental vehicles travel; (4) the fire insurance industry and its requirements; (5) the deterioration, flammability, and conditions of properties; and (6) the level of carelessness of the population. Each variable is successively less amenable to control by local fire department officials. Yet, hold the first variable constant, and there still may be wide variation in the performance of the urban fire protection function. The first difficulty of service analysis in vacuo is that it ignores the constraints imposed by, and the costs borne by, the private sector in the provision of a social goal.

The role of the nonpublic sector is also important because there are differences in the degree to which the private sector satisfies the needs and tastes of different social groupings. Public parks and pools may be more important in neighborhoods without spacious lawns and backyard pools; fire protection may be more crucial in older, denser areas; libraries may be more essential in neighborhoods where private book-buying is a luxury. If the private sector satisfies some social function, is it necessary to insist, just for the sake of equality, that the public sector should perform it as well? Alternatively, if the private sector is not serving some need, is it simply because some citizens do not value it very highly? Avoiding the fallacy of partial analysis does not require that analyst examine the total package of forces which contribute to the satisfaction of some public function. To do so would be a heroic feat. But we should not confuse service output with goal-attainment.

THE ECOLOGICAL FALLACY

To hold that statements about individuals cannot be accurately inferred from data about sociospatial units is an expression of the ecological fallacy. With sociospatial data, it will be impossible to assert that "rich citizens benefit more than poor ones," or that "black citizens are burdened more than white citizens." Rather, data on sociospatial units warrant inferences only about the characteristics of the units. Sometimes, service outputs can be disaggregated to the household level. Even where this is possible, it is not always desirable. In the first place, the character of a neighborhood is a crucial sociopolitical fact to politicians and to households contemplating relocation. Politicians, planners, and administrators simplify reality by aggregating their perceptions to neighborhood levels, a tendency exacerbated by a ward system. As one ward councilman said to another: "You bastard, you had three more blocks of blacktopping in your ward last year than I had, and you'll not get another vote from me until I get three extra blocks."[26] Moreover, some public service outputs cannot easily be disaggregated to households. Such outputs as police deployment practices, urban renewal projects, the location of a sewage treatment plant, and fire station locations are not "delivered" to households *per se,* but to grosser sociospatial units. In any case, so long as we confine our inferences to statements about areal units, we shall avoid the ecological fallacy.

THE PROBLEM OF PUBLIC GOODS

Some outputs of government are *pure public goods.* (The adjective "pure" distinguishes these outputs from the more general phenomena of "public goods and services.") By definition, pure public goods cannot be differentially distributed because no divisibility principle is available. If Citizen A benefits, so does Citizen B. James Buchanan's example of a lighthouse as a pure public good should suffice as an illustration. Samuelson's classic paper on public goods begins by "explicitly assuming two categories of goods." The first is "a *private* consumption good, like bread, whose total can be parcelled out among two or more persons, with one man having a loaf less if another gets a loaf more." The second is a "*public* consumption good like an outdoor circus or national defense, which is provided for each person to enjoy or not, according to his tastes."[27]

Samuelson's initial formulation has engendered much criticism and elaboration,[28] which need not concern us here. Obviously, to the degree that governmental outputs are purely public goods, their distribution to races, classes, and neighborhoods is theoretically impossible. But, while there are some pure public goods produced by urban governments—pollution control, for example— we may agree with Kevin Cox that

> the concept of a public good is in many ways an abstraction from reality. Most public goods have some private aspect to them in terms of restricted availability or exclusion. Such public goods are referred to as "impure public goods." Although no one in a municipality can be excluded from the use of a municipal fire service, there may be variations in availability depending upon distance from the fire station. Moreover, there may be exclusion rules for the consumption of public goods.[29]

By definition, then, we cannot depict the distribution of pure public goods. Most urban services, though, probably fall into the domain of "impure public goods."

EXTERNALITIES

All public policies entail externalities or spillover effects. Some are immediate and apparent; others resemble the radiation which can drift for miles and settle upon an object unobserved until mutations occur years later. Specifying the externalities of any policy distribution would involve the analyst in endless complications. The placement of recreational facilities may affect the use of leisure time, which may affect numerous other variables. Such linkages are too complex to explore at this point, as we shall have our hands full simply specifying immediate distributions of policy, let alone their external effects.

Methods and Measurement

Though we are cognizant of these barriers to analyzing urban services, we can surmount enough of them to make the effort worthwhile. Indeed, the next two chapters will show that some fairly definitive answers can be given to the question of whether services are distributed unequally and why their distribution fits one or another hypothesis. We now have in hand a case in point and five possible hypotheses about service distribution. Four of these are readily subject to empirical examination across a range of neighborhoods. Because the decision-rule hypothesis cannot so readily be quantified, its treatment will have to be qualitative rather than quantitative. In general, if we can measure selected services, identify relevant neighborhood traits, and relate the two, we should be able to assess patterns of service distribution. Specifically, we should be able to answer two questions: (1) who gets the benefits? and (2) who pays the bills?

UNITS OF ANALYSIS

For lack of a better source of classification of neighborhood, we rely upon the familiar census tract as a surrogate. In doing so, we follow the lead of a generation of social scientists studying cities. None of them would make the census tract their first choice for a definition of neighborhood, but as a unit of measure it offers some well-established virtues. Not the least of these is the range of variables available from the decennial census. In San Antonio, at least, census tracts represent rather homogeneous sociospatial entities and exhibit a very wide range of variation on most independent and dependent variables. A total of 113 will be available for our analysis, certainly a large enough number to permit some statistical manipulation.[30]

The array of data which one might in principle secure on census tracts is enormous. The census bureau, the standard repository of this multitude of facts and figures, records information on census tracts ranging from schooling levels to family composition to appliance ownership. Because we can draw upon our hypotheses to select the most useful indicators, we are concerned only with identifying certain key measures of socioeconomic, ethnic, and ecological aspects of these 113 census tracts. In addition, we are interested in developing sufficient data to permit a test of the political power hypothesis, a matter on which the census is naturally silent. The twenty independent variables listed in Table 3.4 represent the range of indicators selected to examine the underclass and ecological hypotheses. None of the measures of socioeconomic status should strike even the casual user of the census as unusual. The measures of racial composition are equally straightforward, being merely measures of the proportion of the two dominant minority groups in San Antonio, Chicanos and blacks, and a sum of the two to indicate the total minority proportion in each tract. The two ecological measures show crucial features of the character of the neighborhood itself, its density and its age.

The measures of the power ascription of a neighborhood are more complex. Even though the data on a social area's political power may be as crucial a fact as its income, education, or occupancy rates, no public data are ever found on the power dimension. It does not, however, defy common sense to believe that political power is spatially distributed, and that some neighborhoods have more of it than others. When a zoning change threatens, when a need is articulated for more police protection or for a new school or traffic signal, or whatever the problem, some areas pack more punch. Parenti, for example, describes the efforts of citizens in a black ward of Newark to secure a traffic light. Three hundred and fifty signatures and three years of petitioning and pressuring failed to secure the light, while a nearby white, middle-class neighborhood was able to get a light installed within twenty-eight days of submitting about fifty signatures on a petition.

Areal political power will depend upon many things—support for the winning slate, degrees of political organization, level of participation, and the like. But if there is any single good index to an area's political muscle, it will likely be suggested by the number of formal and informal public offices held by neighborhood residents. To be sure, even neighborhoods without any "representation" in the councils of power may get attention paid to their needs. But the unrepresented group is always at the mercy of someone else's good will. The concentrated nature of political power in San Antonio should make it relatively easy to measure political influence.

To be certain that we have not focused merely on one or two possible positional roles, we triangulated on power by including a battery of measures, consistent with Etzioni's dictum that "any single measure of a phenomenon is suspect." The roles identified included council member, member of the board of directors of the Good Government League, and municipal department heads. Both raw and ratio measures are incorporated, and a longitudinal dimension is introduced by cumulating the total number of council members from 1954 (at the inception of the GGL) to 1972. After all, public service facilities have a certain permanence about them, and accumulated power may be more important than immediate power. Collectively, these measures offer a spatial perspective on political power, and enable us to ascertain whether an area's residential power base is associated with the level of its services.

Table 3.5 lists for every independent variable the mean, standard deviation, range, and coefficient of variation, the latter measuring the degree to which any trait is dispersed among the population. Almost all of these independent variables show a great range of variation among the census tracts of San Antonio. Some tracts are virtually all black, some have no blacks; some tracts have nearly two-thirds of their residents at or below the poverty level, others have none; a good many tracts have had no member of the city council in the period 1954-1972, but one had as many as seventeen. This range of variation at once tells us a good deal about inequalities in San Antonio. The coefficient of variation is a succinct and useful measure of inequalities in these attributes. The

Table 3.4: Independent Variables

Name	Definition and Measurement
I. Socioeconomic status	
1. Median school years	Median school years completed by persons 25 years old and over
2. Median family income	Median income of families
3. Percentage poverty families	Percentage of all families with income at or below census-defined poverty level
4. Percentage owner occupancy	Percentage of housing units occupied by owner
5. Percentage white collar	Percentage of workers in white collar occupations
6. Median gross rent	Median rent for renter-occupied housing units.
7. Percentage overcrowded housing	Percentage of housing units with more than 1.01 persons per room
II. Racial composition	
1. Percentage Negro	Percentage nonwhite
2. Percentage Spanish heritage	Percentage Spanish heritage
3. Percentage minority	Sum of percentage Negro and percentage Spanish heritage
III. Ecological characteristics	
1. Age of neighborhood	Percentage of year-round housing built in 1949 or earlier
2. Density	Population per acre
IV. Political power	
1. Number of council members	Total number San Antonio City Council members residing in tract 1955-1972
2. Ratio of council members to population	Ratio of total number of council members residing in tract, 1955-1972, to population of tract
3. Number of top bureaucrats	Number of municipal department heads residing in tract in 1972
4. Ratio of top bureaucrats to population	Ratio of number of resident department heads to population
5. Number of electoral elites	Number of GGL directors and city council members residing in tract in 1972
6. Ratio of electoral elites	Ratio of resident GGL directors and city council members to population
7. Number of political elites	Number of municipal department heads, GGL directors, and city council members residing in tract in 1972
8. Ratio of political elites to population	Ratio of resident municipal department heads, GGL directors, and city council members to population

Table 3.5: Independent Variables: Means Standard Deviations, Ranges,
           and Coefficients of Variation

	Variable	X	s	CV	Range
I.	*Socioeconomic status*				
	1. Median school years	9.96	2.42	.24	13.9 − 4.9
	2. Median family income	$7,786	$2,870	.36	$17,214 − $2,347
	3. Percentage families	18.48	13.35	.66	67.5 − 1.6
	4. Median value of owner-occupied dwelling units	$12,031	$5,390	.45	$28,500 − $6,100
	5. Percentage owner occupancy	59.57	18.39	.31	96.60 − 5.20
	6. Percentage white collar	47.72	19.96	.41	89.1 − 17.9
	7. Percentage over-crowded housing	17.36	12.49	.71	47.6 − 1.8
II.	*Racial composition*				
	1. Percentage Negro	8.45	20.64	2.44	95.9 − 0.0
	2. Percentage Spanish heritage	50.99	30.35	.58	99.0 − 4.7
	3. Percentage minority	60.36	31.90	.53	100.0 − 5.8
III.	*Ecological characteristics*				
	1. Age of neighborhood	46.60	32.25	.69	95.3 − 0.0
	2. Density	8.22	20.64	.69	26.1 − 1.0
IV.	*Political power*				
	1. Number of council members	.79	2.22	2.80	17.0 − 0.0
	2. Ratio of council members to population	.13	.35	2.69	2.17 − 0.0
	3. Number of top bureaucrats	.67	1.35	2.01	6.00 − 0.00
	4. Ratio of top bureaucrats to population	.14	.29	2.07	1.96 − 0.00
	5. Number of electoral elites	1.99	4.01	7.02	23.00 − 0.00
	6. Ratio of electoral elites to population	.32	.58	1.81	3.03 − 0.00
	7. Number of political elites	2.66	4.77	1.79	28.00 − 0.00
	8. Ratio of political elites to population	.45	.71	1.58	3.69 − 0.00

larger it becomes, the greater the variation or inequality among neighborhoods with respect to any particular attribute. The most extreme inequalities across the board are found in the measures of political power. Whether they are measured on a raw or ratio basis, the political power variables show wide disparities from neighborhood to neighborhood. This certainly squares with our verbal description above of the nature of power concentration in San Antonio. It seems clear that generally political inequality is much sharper and more widespread than socioeconomic and even ethnic inequalities.

The only other independent variable approximating such disparities from neighborhood to neighborhood is the percentage of blacks in the tract. There are a very few census tracts (four) with more than three-quarters of their population black, and a very large number (seventy-four) with no blacks at all. In comparison to the high level of segregation between blacks and other ethnic groups, Chicanos are relatively more dispersed.

URBAN SERVICES: MEASURING MUNICIPAL OUTPUTS

As long ago as 1938, Ridley and Simon wrote a well-before-its-time monograph on *Measuring Municipal Activities*.[32] They argued for "efficiency" as the proper criterion for urban administration and spelled out suggestions for measuring efficiency in numerous areas. No one took up these suggestions, nor the more general question of assessing urban service outputs, for a very long time. Only in the 1960s did social scientists and public administrators pay much attention to the problems initially raised by Ridley and Simon. Even then, much of the discussion was parameterological, as many wrote of the need to measure service outputs, but few did anything about it. Actually, in the 1960s, the focus of most urban political analysis was on process, participation, and power, rather than upon policy. Urban political analysts became very adept at explaining the decisional process, but tended to ignore the decisions themselves. Dahl's "decisional approach" to power structure used decisions to identify power holders, but the decisions themselves were almost incidental to his analysis.[33] The tail of process came to wag the dog of policy. In terms of Lasswell's famous aphorism about politics as "who gets what how," less was known about the "*what* was gotten" than the "*how* it was gotten" part of the aphorism.

The emergence of "public policy analysis" in political science partly rectified some of these deficiencies by focusing on the "outputs" of the political system. In the literature of both state and urban politics, a new brand of research concerns emerged, the major thrust of which was to identify the socioeconomic and political correlates of policy variations. These developments were generally constructive, if only because they emphasized that political scientists are concerned with what governments do, and not merely with how things get done. But the policy outputs research failed to confront the full implications of the Lasswellian paradigm. It focused almost entirely upon the *levels* of public policy, typically measured in terms of dollars and cents. Its emphasis on "who spends what how" really sidestepped the distributional and allocational implications of Lasswell. If one of the objects of policy analysis is to develop and test hypotheses about the benefits and burdens of policies, then it is not enough to know how big the policy pie is. It is at least equally important to know how it is carved up.

Fortunately, the last several years have seen some action on the front of service measurement. Particularly, this is true in education, but also in the areas of the more traditional urban services like parks, roads, libraries, police protection, and the like, both pure and applied work on service measurement has begun to appear with regularity.

This research has been undertaken by the Urban Institute,[34] the RAND Corporation,[35] and others.[36] A good part of it is reviewed when we investigate service outputs and their distribution in the next two chapters. The array of public services which can be and have been investigated is large indeed. Unlike the American states, which spend most of their resources on the "big three" items of education, welfare, and highways, urban expenditures are disaggregated into numerous categories. Among the traditional classifications of city government expenditures—education (which only a few *municipalities* operate), parks, roads, libraries, hospitals, sanitation, fire, airports, police, and so forth—no one function accounts for as much as 10 percent of the total expenditures. San Antonio spends nearly half a million dollars on the relatively minor function of brush removal. So out of the great range of services, we must make some choices.

For our purposes, we focus on measures both of burdens and benefits from urban government. We have only one measure of burden—the estimated ratio of property assessment to true value— but a multiplicity of service output indicia. All of these are listed in Table 3.6. Six major service areas of local government are represented there: police, fire protection, parks and recreation, public water supply, public sewers, and libraries. Collectively, these measures focus on the distance, quality, and consumption aspects of urban service allocation. For some public facilities, particularly parks, fire protection, and libraries, proximity will be

Table 3.6: Dependent Variables: Measures of Service Outputs to Census Tracts, San Antonio, Texas

Service	Measures
I. Parks and recreation	1. Park distance
	2. Developed acreage at closest park
	3. General evaluation of closest park
	4. Pool distance
	5. Pool quality
	6. Playground quality
	7. Sportsfield quality
	8. Playground use
	9. Sportsfield use
	10. Pool use
II. Police patrol	1. Man-units of police patrol to census tract during calendar year
III. Fire protection	1. Fire station distance
IV. Libraries	1. Library distance
	2. Library volumes per capita
	3. Library personnel per 1,000 population
	4. New books per 1,000 population
	5. Library expenditures per capita
V. Public sewers	1. Percentage housing units with public sewers
VI. Public water	1. Percentage housing unit with public water
VII. Tax evaluation	1. Ratio of tax assessment to estimated value of housing unit

a crucial determinant of utilization. Especially for parks and libraries, we have measures of relative quality of services as well as indicators of proximity. Because all of the service measures in Table 3.6 can be coded to particular census tracts, we will be able to relate the attributes of those tracts to service outputs, exploring the underclass hypothesis and the ecological hypothesis. Each of the indicia will be more precisely defined when we analyze service distributions in Chapter 5, but the overview here should suffice to provide an acquaintance with the measures.

These service indicia hardly exhaust the range of possibilities. Nor will they overcome the various difficulties in service output analysis we identified earlier in this chapter. To the contrary, they offer only a first cut. They permit some empirical tests of the several hypotheses with a range of data on a multiplicity of service indicia. What that analysis tells us about equality in the urban public sector is the subject of the next two chapters.

NOTES

1. Major annexations in the early 1970s moved San Antonio from the rank of fifteenth largest city in 1970 to an estimated rank of eleventh in city size.

2. Herein we follow the useful southwestern convention of referring to all non-black and non-Chicano persons as "Anglo." According to the 1970 census, the proportions of San Antonio population for each group were: Anglo—40.2 percent; black—7.6 percent, and Chicano—52.1 percent. Almost surely, such figures understate the number of Chicanos in the city, and their numbers are increasing at a faster rate than those of other groups.

3. In the April, 1973 municipal elections, the GGL was able to secure only a five-four majority on the city council and lost the mayoral post. By no stretch of the imagination, however, could the challengers to the old hegemony be described as more liberal than the GGL.

4. The best discussion of the GGL is found in Truett Chance, "The Relationship of Selected City Government Services to Socioeconomic Characteristics of Census Tracts in San Antonio, Texas," unpublished Ph.D. dissertation, Department of Government, University of Texas at Austin, 1970, pp. 24-34.

5. See, for example, Robert L. Lineberry and Edmund P. Fowler, "Reformism and Public Policies in American Cities," *American Political Science Review,* 61 (September, 1967), pp. 701-716; and Edward C. Banfield and James Q. Wilson, *City Politics* (Cambridge, Mass.: Harvard Univ. Press and MIT Press, 1963).

6. Theodore Lowi, *The End of Liberalism* (New York: W. W. Norton, 1969), pp. 44-45.

7. Charles M. Tiebout, "A Pure Theory of Local Expenditures," *Journal of Political Economy,* 64 (October, 1956), p. 190.

8. The source of the Austin case is Steven J. Kraus, "Water, Streets, and Sewers: The Acquisition of Public Utilities in Austin, Texas, 1895-1930," unpublished M.A. thesis, Department of History, University of Texas at Austin, 1973.

9. Cited *ibid.,* pp. 150-151.

10. Patricia Cayo Sexton, *Education and Income* (New York: Viking, 1961), p. 16.

11. See our discussion in Chapter 1.

12. Floyd Hunter, *Community Power Structure* (Garden City, N.Y.: Doubleday Anchor, 1953), pp. 18 and 21.

13. Roger Kasperson, "Toward a Geography of Urban Politics: Chicago, a Case Study," *Economic Geography,* 11 (April, 1965), pp. 95-107.

14. Francis Rourke, *Bureaucracy, Politics and Public Policy* (Boston: Little, Brown, 1969), p. 60.

15. Ira Sharkansky, *The Routines of Politics* (New York: Van Nostrand Reinhold, 1970).

16. Otto Davis, M. A. H. Dempster, and Aaron Wildavsky, "A Theory of the Budgetary Process," *American Political Science Review,* 60 (September, 1966), pp. 529-547; John P. Crecine, *Governmental Problem Solving: A Computer Simulation of Municipal Budgeting* (Chicago: Rand McNally, 1969); and Sharkansky, *op. cit.*

17. John A. Gardiner, "Police Enforcement of Traffic Laws," and Martha Derthick, "Intercity Differences in Administration of the Public Assistance Program: The Case of

Massachusetts," both in James Q. Wilson, ed., *City Politics and Public Policy* (New York: John Wiley, 1968), pp. 151-172 and 243-266. On discretion in public agencies more generally, perhaps the best treatment is Kenneth C. Davis, *Discretionary Justice* (Baton Rouge: Louisiana State Univ. Press, 1969).

18. James Q. Wilson, *Varieties of Police Behavior* (Cambridge, Mass.: Harvard Univ. Press, 1968), p. 30.

19. See his paper, "Toward a Theory of Street-Level Bureaucracy," in Willis D. Hawley, et al., *Theoretical Perspectives on Urban Politics* (Englewood Cliffs, N.J.: Prentice-Hall, 1976), Chap. 8.

20. Werner Z. Hirsch, "The Supply of Urban Public Services," in Harvey S. Perloff and Lowden Wingo, Jr., eds., *Issues in Urban Economics* (Baltimore: Johns Hopkins, 1968), p. 519.

21. Carl S. Shoup, "Standards for Distributing a Free Government Service: Crime Prevention," *Public Finance,* 19 (1968), p. 383.

22. Andrew Boots, et al., *Inequality in Local Government Services: A Case Study of Neighborhood Roads* (Washington, D.C.: The Urban Institute, 1972).

23. See, e.g., the discussion and critiques in Frederick J. Mosteller and Daniel P. Moynihan, eds., *On Equality of Educational Opportunity* (Cambridge, Mass.: Harvard Univ. Press, 1972).

24. For a more extended inventory of urban service indicators, see Robert L. Lineberry and Robert E. Welch, Jr., "Who Gets What: Measuring the Distribution of Urban Public Services," *Social Science Quarterly,* 54 (March, 1974), pp. 700-712.

25. Elinor Ostrom, "On the Meaning and Measurement of Output and Efficiency in the Production of Urban Police Services," *Journal of Criminal Justice,* 1 (June, 1973), p. 97.

26. Oliver P. Williams and Charles Adrian, *Four Cities* (Philadelphia: Univ. of Pennsylvania Press, 1967), p. 264.

27. Paul Samuelson, "Diagramatic Exposition of a Theory of Public Expenditures," *Review of Economics and Statistics.* 37 (November, 1955), p. 350.

28. Julius Margolis, "A Comment on the Pure Theory of Public Expenditures," *Review of Economics and Statistics,* 37 (November, 1955), pp. 347-349; Norman Frolich and Joe A. Oppenheimer, "A Reformulation of the Collective Good-Private Good Distinction," a paper presented at the annual meeting of the Public Choice Society, Pittsburgh, Pa., May 3-6, 1972.

29. Kevin Cox, *Conflict, Power, and Politics in the City: A Geographic View* (New York: McGraw-Hill, 1973), p. 25.

30. We have excluded such tracts as do not lie sufficiently within the boundaries of San Antonio to permit confident inclusion of their characteristics, and have also excluded tracts which are wholly coterminus with the several military bases in the city, leaving a total of 113 tracts for analysis.

31. Michael Parenti, "Power and Pluralism: A View from the Bottom," *Journal of Politics,* 32 (1970), pp. 512-513. More generally on the efforts of residents of disadvantaged neighborhood to maximize their political power, see Norman I. Fainstein and Susan S. Fainstein, *Urban Political Movements: The Search for Power by Minority Groups in American Cities* (Englewood Cliffs, N.J.: Prentice-Hall, 1974).

32. Clarence E. Ridley and Herbert A. Simon, *Measuring Municipal Activities,* (Chicago: International City Managers' Association, 1938).

33. Robert A. Dahl, *Who Governs?* (New Haven, Conn.: Yale Univ. Press, 1961).

34. See, e.g., Peter B. Bloch, *Equality in the Distribution of Police Services: A Case Study of Washington, D.C.* (Washington, D.C.: The Urban Institute, 1974); Louis H.

Blair and Alfred I. Schwartz, *How Clean is Our City?* (Washington, D.C.: The Urban Institute, 1972); and Boots, et al., *op cit.*

35. See, e.g., Robert K. Yin, "On the Equality of Municipal Service Outcomes: Street Cleanliness," paper presented to the Annual Meeting of the Operations Research Society of America, April, 1974.

36. Frank S. Levy, et al., *Urban Outcomes: Schools, Streets, and Libraries* (Berkeley: Univ. of California Press, 1974); John C. Weicher, "The Allocation of Police Protection by Income Class," *Urban Studies,* 8 (October, 1971), pp. 207-220; and Charles S. Benson and Peter B. Lund, *Neighborhood Distribution of Local Public Services,* (Berkeley: Institute of Governmental Studies, University of California, 1969).

Chapter 4

WHO BEARS THE BURDENS?

But in this world, nothing is certain but death and taxes.

Benjamin Franklin

The corruption of democracies proceeds directly from the fact that one class imposes the taxes and another class pays them.

Dean Inge

It may be said that nothing is so certain as death and taxes, but that death is more evenhanded. The issues of fairness and equity in taxation have been around for decades, periodically rearing their heads in election campaigns, and then often retiring to obscurity amid the ebb and flow of more immediate political issues. Yet, at the urban level, taxes are almost always a more acute political problem than at the state or national levels. For one thing, local taxes strike—quite literally—closer to home. The property tax has long been the mainstay of local governmental finance, and there is no tax upon which citizens and specialists are more likely to agree in their condemnation. Almost every discussion of equity and equality in urban governance will ultimately come round to a problem of the urban property tax. To know something about tax assessments is to know something about one of our two fundamental questions: Who bears the burdens?

The Tax Assessment Issue

In the fall of 1973, when the overall state of American government was quite distressing to most Americans, the United States

Senate Subcommittee on Intergovernmental Relations commissioned Louis Harris and Associates to survey the American population concerning their views of American institutions. The general state of the public mind was not very comforting. Only medicine and local trash collection received the unqualified support of a majority of Americans. The remaining institutions in the survey trailed well behind, down to 30 percent support for the Senate itself, 29 percent for major companies, and 24 percent for state government. Dragging up the rear, and not surprisingly so in the context of the events of the period, were the federal executive branch and the White House. The only institution to tie the White House for dead last (with support scores of 19 percent) was local tax assessment.[1]

There is a myriad of reasons for the low esteem in which the local tax assessment system is held, far too many to permit even a review here. To the tax economist, the property tax is characterized by its inelasticity, by its enormously expensive collection costs, and by its perpetuation of taxation on a relatively outmoded form of wealth in the nation. The citizen is likely to have an even grimmer view of the assessment system than the tax specialist. Almost no one is fond of the local tax assessor. Even the normally restrained Advisory Commission on Intergovernmental Relations, in describing the property tax assessment system as one of the few "treasured relics of pioneer days" that cannot be found in a museum, added that "the average assessor makes himself a sort of one-man legislature."[2] Even though there may be appeal structures—sometimes labyrinthine ones—available to the aggrieved taxpayer, courts are especially loathe to intervene in local tax disputes and have given the widest possible discretion to local tax assessors and collectors in almost every state. Hearing every nickel-and-dime tax challenge could further overtax an already burdened court structure.

Being subject to the widest possible range of administrative discretion, further beclouded by complex and frequently hidden formulae, rules, and precedents, local tax administration is subject to the most imaginative array of corruption to be found in the American system of government. No doubt for a variety of reasons, the property tax has come to represent one of the rudest intrusions of government into the lives and fortunes of its citizenry. Perhaps more because of the property tax than because of the character of the schools themselves, the year 1971 was the first in which a majority of school bond referenda were defeated by the electorate.

In Texas, the property tax is not much different from anywhere else, save for the fact that San Antonio was the site of the major challenge to the property tax-based funding of the public schools. In a brief submitted in "San Antonio Independent School District v. Rodriguez," Professor Joel Berke of Syracuse University demonstrated that the property tax structure tended in the state of Texas generally, and Bexar County particularly, to key school resources to local wealth.[3] Curiously, the United States Supreme Court seemed not only to ignore Berke's evidence in its majority opinion, but it strayed to some data from Connecticut showing that the property tax system there did not necessarily disadvantage poor families.[4] School districts and cities, however, are different things. Even though the Bexar County school system is extremely fragmented (from the dirt-poor Edgewood district, in which Dimitrio Rodriguez lived, to the prosperous Alamo Heights district), the City of San Antonio is relatively centralized for tax collection purposes. It is true that tax assessment ratios will differ from city to school district to county, even though the same office handles assessment and collection. But the City of San Antonio is our sole focus herein, and its assessment practices are the subject of our inquiry.

The San Antonio municipal government operates within the context of an elaborate and antiquated system of rules, statutes, and court decisions concerning assessment practices. As one observer comments, "the pattern of property tax administration in Texas is no exception to the national picture. Texas provides a model of inept and unfair administration. . . . Despite the widespread and informed criticisms of the Texas property tax, both the legislature and the state courts have been unresponsive to the necessity for reform. While other states have moved to correct property tax abuses and inequities, Texas decision-makers have chosen to remain silent."[5]

As with most states and cities, the burden of the assessor is an impossible one. Under the law, it is the taxpayer's responsibility to "render" his property—personal as well as real—for taxation rolls. There is, of course, little compulsion for the taxpayer to render his property at its fair market value, for to do so would almost certainly set his tax rate at higher than others similarly situated. Assessment thus becomes, for all practical purposes, a matter of negotiation between the tax assessor and the taxpayer. Home-owners almost invariably accept the assessor's evaluation and rarely avail themselves of the opportunity to run the gamut of appeals. When they do, they are infrequently successful, for the equalization board almost never upholds a challenge to an assessment.[6] The only exceptions typically concern an aspect of the property (e.g., poor drainage) of which the tax office was initially unaware.

The ambiguities and resultant inequities of tax assessment in Texas arise, however, not so much from the studied discrimination of the tax office as from the literal impossibility of accomplishing lawful requirements. Theoretically, all property is assessed on June 1 of the year. Practically, there is no way to assess everything on a single date. Theoretically, all property—personal as well as real—is to be rendered, a requirement long fallen into disuse because of its impracticality. Theoretically, equivalent properties are assessed at equivalent values. Yet, the range of assessments in Texas both between and within particular classes of property varies enormously.[7] Theoretically, there are formulae available for determination of property values, including everything from the value of a fireplace to the value of a square foot. Yet, in San Antonio, at the time of the present study, the "cost analysis" book was last updated in 1953. It would be remarkable if assessment practices in San Antonio did not exhibit inequalities, and entirely possible that such inequalities are systematically related to attributes of various neighborhoods.

Unquestionably, the burden of other evidence suggests that tax assessment practices vary as widely within jurisdictions as they do between jurisdictions. The Advisory Commission on Intergovernmental Relations developed a minimum level of acceptability for equalization patterns, and discovered that almost 40 percent of large assessment districts had coefficients of dispersion which fell beyond the acceptable range.[8] Such general arguments are supported by studies of individual taxing jurisdictions. Perhaps the most useful of these studies are the so-called "ratio studies," in which actual sales prices of units are compared to the unit's assessment. The result is a ratio representing the degree to which the assessed value corresponds to the sale value. Oldman and Aaron, for example, developed a rich set of data on assessments in Boston, Massachusetts, including data on the prices of 13,769 sales in the early 1960s.[9] They discovered variations both between and within different types of property. For 1962, single family residences were assessed at the lowest levels (only 34 percent) while commercial property was assessed at the highest value (79 percent). There were also variations within different classes. There was considerable deviation in the assessments of commercial properties, and also among residential properties of different values. In Boston, at least, the effect of the property tax is highly regressive. Homes of relatively low value were assessed at much higher ratios than homes of relatively high value. Such evidence seems to confirm the conventional economic wisdom which holds that the property tax is essentially a regressive tax.[10]

At the same time, it is worth noting that San Antonio is not Boston. One would not confuse Old North Church and the Alamo, and one should not confuse an old, multiethnic city hemmed in by its suburbs, completely built up, with a new, still-growing, vigorously annexing, and triethnic city like San Antonio. Even if tax assessors may be all alike, cities are not. What we propose to do, therefore, is to test the proposition that the burdens of local government are borne differentially by different sociospatial groups within San Antonio. Our unit of analysis, it will be recalled, is areal, not individual. Operationally, we are concerned with variations in assessment practices amongst census tracts, not from house-to-house and store-to-store. What we shall do, after explaining our measure of the dependent variable of assessment burden, is to explore whether the burdens of local government are systematically related to those variables identified by our major hypotheses: race, socioeconomic status, political power, and ecological attributes of the neighborhood. But first, however, it will be useful to identify some problems in the measurement of the dependent variable.

Tax Burdens: The Measurement Problem

If one resorts to the legal concept of "best evidence," the soundest analysis of assessment practices would involve some replication of the "ratio studies" described above. Such studies require two kinds of data. First, data on some sample or universe of property sales is secured, typically because the government requires a stamp affixed to a deed on which a selling price is specified. Such information then becomes available from government sources who dispense sales stamps. There are other vehicles for securing sales information, the most obvious of which is the local real estate community. The second item of data needed is the assessment assigned to a particular parcel of property. The assessment data is a straightforward item normally available at the tax office. The combination of these two data for every parcel produces a sales ratio for each piece of property, which can then be aggregated in whatever way desired, by neighborhood, census tract, type of property (commercial v. industrial v. vacant land v. multifamily v. single family) and so on.

When the ideal is elusive, however, one makes one's peace with the real. For the period of our study, there was no practical way to secure data on sales prices. As stamps were not required at that time, and multiple listing services secreted their own data from tax authorities, we fell back on a strategy of indirectly estimating residential property values. We secured a one-half percent sample of the 224,893 properties on the 1972 tax rolls of San Antonio, replacing all nonsingle family dwelling units in the sample with the next single-family residential property. Each of these properties was coded by the census tract in which it was located and for its assessed valuation. That half of the enterprise was fully consistent with the assumption of the ratio study. The construction of the second half of the ratio, however, required a reliance on the Census Bureau's estimate of the median value of owner-occupied homes in the census tract. The latter is our substitute for the estimate of value derived from the sale of a property, and is admittedly an ersatz alternative. The Census Bureau relies exclusively on respondent estimates of value (as, of course, they do on such items as age and income). Obviously people differ in the accuracy of their estimates. Whether such errors of estimate are systematic or random, no one could say. It would be safe to say, however, that such estimates, *taken in the aggregate, i.e., for census tracts as a whole,* bear a very distinct relationship to common knowledge evaluations of property values in different neighborhoods. For that reason, they have been used by countless sociologists, political scientists, economists, and market analysts, as surrogates for the socioeconomic status of the census tract or neighborhood.

The resulting measure, therefore, in the present study, is the mean value of the sampled assessments for census tract X as a percentage of the census figure on the median value of owner-occupied dwelling units for census tract X. The first figure is derived from approximately ten single family dwelling units per tract which fell into our sample. The closer the ratio of the two numbers is to 1.00, therefore, the more exactly assessed value in the tract approximates "true" value as herein estimated. The larger the ratio, the lower is the assessment in relationship to the "true" value of housing in that tract.

The mean value of our assessment ratio, thus constructed, is 33.48 percent, indicating that the mean census tract contains an assessment of approximately one-third of its census-identified value. Having this measure in hand, we can now proceed to examine what—if any—relationship it bears to our hypothesized correlates. Specifically, in accord with the underclass hypothesis, we suggest that

1. Census tracts with higher socioeconomic status exhibit lower assessed values in relation to true values;

2. Census tracts with larger proportions of minorities exhibit higher assessed values in relation to true values; and

3. Census tracts with greater political power exhibit lower assessed values in relation to true values.

Once we examine the evidence for these hypotheses, we can then proceed to determine whether the ecological attributes of neighborhoods—their newness, their density, and the like—contribute to variations in their assessed valuation. Finally, we can provide a multivariate analysis which asks, essentially, which neighborhood

attributes in San Antonio contribute most significantly to variations in the assessment practices of city government.

The Underclass Hypothesis and Local Taxation

CLASS AND RACE

As we have already suggested, the conventional wisdom about
urban assessment practices holds that the poor and the minorities,
whatever minority they may represent, bear the brunt of discriminatory practices. This hypothesis is even borne out in some of the
assessment studies noted above. By some simple statistical analysis, we can test this hypothesis in San Antonio. Table 4.1 represents a first effort. There we report simple correlation coefficients
between various indicia of socioeconomic status and ethnicity on

Table 4.1: Correlations between Socioeconomic Status, Racial Composition,
and Political Power of Neighborhoods and Tax Assessment Ratios,
San Antonio, Texas*

Independent Variables	r
I. *Socioeconomic status*	
1. Median school years	.493
2. Median family income	.520
3. Percentage poverty families	−.540
4. Median value of owner-occupied dwelling units	.447
5. Median gross rent	.478
6. Percentage white collar	.539
7. Percentage overcrowded housing	−.408
II. *Racial composition*	
1. Percentage Negro	−.338
2. Percentage Spanish heritage	−.385
3. Percentage minority	−.588
III. *Political power*	
1. Number of electoral elites	.238
2. Ratio electoral elites to population	.229
3. Number political elites	.316
4. Ratio of political elites to population	.316
IV. *Ecological characteristics*	
1. Population mobility	−.322
2. Age of neighborhood	−.433
3. Density	−.517

*All the relationships reported herein are significant at the .01 level.

the one hand and the dependent variable on the other. It is important to note that a positive sign represents a direct relationship between the value of the independent variable and the size of the assessment ratio. Thus, a high positive correlation will show that the increase in one attribute (say, median income) is associated with a closer approximation of assessed to true value.

Indeed, what Table 4.1 shows is surprisingly unambiguous. The *higher* the socioeconomic status of a neighborhood, the *greater* its assessment in relationship to the census-derived value of housing. Similarly, the *higher* the proportion of Anglos in a neighborhood, the *greater* is the assessment of property in that neighborhood. While none of these correlations are so strikingly high as to settle definitively the question, they do exhibit the virtue of absolute consistency, a consistency bolstered by an acceptable level of statistical significance. However measured, the correlations in Table 4.1 suggest the rejection of hypotheses 1 and 2 concerning class and racial explanations of tax assessment practices. In fact, the strongest single correlates are the indicia of poverty and percent minority. In each of those cases, the presence of considerable poverty, or large numbers of minority citizens, is associated with lower assessments. Of the eighteen census tracts with 10 or more percent Negro, for example, only three show an assessment score as large as the city's average.

This does not show, of course, that the property tax is in fact progressive. The data bear only on the assessment, not on the overall burden of the tax. The property tax can be regressive for at least two reasons. The first is because housing expenditures take a larger percentage of the incomes of the poor than of the affluent. Just as sales taxes hit the lower income classes harder because more of their expenditures are subject to the tax, so too property taxes strike harder at those whose housing expenditures are a large proportion of their monthly outlays. The second reason for the regressivity charge has to do with assessment practices, and the frequently ventured hypothesis that the dwelling units of the poor are subject to higher assessments. It is the latter explanation for the regressivity argument which the data here cast doubt upon, not the former one.

It might be objected, incidentally, that the present data do not bear upon the burdens of the poor at all, for the poor are not home-owners. Aside from the obvious response that renters carry the indirect burden of the property tax, there is no reason even to accept the assumption that the lower-income groups in San Antonio are not home-owners. San Antonio is not a city with huddled masses crowded into dingy tenements. It is a city of homes, however humble some of them may be. The poor (obviously not the poorest of the poor, but the near-poor) are purchasing homes just like the affluent. San Antonio, the nation's poorest big city, still had, in the 1970 census, an owner-occupancy ratio exactly equivalent to the national average (59 percent). Clearly, in San Antonio at least, assessments tend to approximate true values more closely when indicia of socioeconomic status are higher, and when minorities are fewer, thus casting doubt on a good deal of conventional wisdom.

POLITICAL POWER

Table 4.1 also provides some data on the correlations between various indicia of neighborhood political power and their assessments. These data are perhaps not as valuable as the earlier data on socioeconomic status and ethnicity, primarily because the independent variables do not exhibit a wide range of variation. Correlation coefficients are more suspect in such cases. For that reason, it may be useful to bolster our argument with some simpler techniques. We can examine in Table 4.2 the means of tracts with various patterns of political power configurations. For the sake of brevity, we show only three comparisons, the mean assessment ratio of tracts with scores of zero and greater than zero for the number of electoral elites residing in them; the assessment scores of tracts where the ratio of the number of council members to population is zero and greater than zero; and, for greater detail, the various scores of the assessment variable for tracts ranging from no councilmanic residential representation to the high of 17 councilmen residing in the tract.

All of the evidence in Table 4.2 points to the conclusion that assessment practices are not strongly tied to the resident political power in the neighborhood. If anything, the obverse is the case. There appears to be a slight tendency for neighborhoods with no resident officeholders to have somewhat lower assessment ratios than those neighborhoods which have sent a local boy to the councils of power. In fact, the highest assessment ratio of all is scored by that single tract which sent seventeen of its own to the city council. If, in the great game of "friends and neighbors" politics, one hopes to secure favorable assessments for one's neighbors, it does not work well in San Antonio. It remains possible that councilmen and other power holders may do well for themselves and for their friends, but not, it appears, for their neighbors.

Ecological Characteristics and Tax Burdens

Assessment burdens are not only a function of the socioeconomic characteristics of the neighborhood, and of the kinds of people who live there. The category of variables we describe as ecological, having to do with the age, density, stability, and mobility of the neighborhood, may also contribute to variations in assessment practices. The correlations reported for ecological characteristics in Table 4.1 bear on these aspects of assessment policy. Hypotheses in this area can readily cut both ways. One might

Table 4.2: Mean Values of Assessment Scores by Selected Indicia of Political Power, San Antonio Census Tracts (N=113)

Indicator	$\overline{X}$
A. *Electoral elites*	
1. Number electoral elites in tract = 0 (59)	31.63
2. Number electoral elites in tract > 0 (54)	35.45
B. *Ratio council members to population*	
1. Ratio of council members in tract to tract population = 0	32.92
2. Ratio of council members in tract to tract population > 0	35.39
C. *Number of council members residing in tract*	
1. All tracts (113)	33.47
2. No council members (88)	32.92
3. 1 council member (8)	37.29
4. 2 council members (4)	31.08
5. 3 council members (4)	32.20
6. 4 council members (3)	35.68
7. 5 council members (1)	36.69
8. 6 council members (2)	40.97
9. 7 council members (1)	35.69
10. 9 council members (1)	28.58
11. 17 council members (1)	44.03

assume that older neighborhoods are overassessed, precisely because their assessments may not change to reflect deteriorating
property values. Conversely, one might just as logically assume
that older neighborhoods are under-assessed because assessment
practices are decades behind time and do not reflect current
market prices. The answer no doubt depends upon subtle processes of changing property values which longitudinal data would
be required to monitor. What the data in Table 4.1 suggest,
though, is that old and densely populated neighborhoods are
relatively under-assessed in relation to census estimates of home
values there. The reason, very probably, has to do with the relative
vigor by which new subdivisions are assessed promptly at values
closely coincident with then-current housing costs. The tax assessor, operating like all bureaucrats with limited resources of time
and money, will of necessity evaluate new properties first, if they
are to be evaluated at all. Older properties can then be brought up
to current standards at a later time, if resources are ever sufficiently free to permit massive reevaluation of older neighborhoods.
In the tax office, as elsewhere, priorities must be set; decision-rules
must accommodate the press of public business. What can wait
until tomorrow will not be done today. And what can wait until
tomorrow are the laborious chores of reevaluating thousands—
there are after all almost a quarter of a million properties on the
San Antonio tax rolls—of older areas.

This happens to be a decision-rule which seems to redound to the
benefit of those who reside in the older areas of town. These are not
always, of course, the poorer areas. (The correlation between the age
of a neighborhood and its poverty level is a positive, but not
overwhelming .445.) They are, in fact, neighborhoods likely to be
populated by both the relatively lower-income groups and by the
very highest-income groups, particularly those families whose
wealth is old and whose status is long-established. It is the rising
middle classes, those whose dreams are fulfilled by a new home in a
new subdivision, whom the tax assessor is most likely to strike.

A Multivariate Approach

Thus far, we have focused on the correlates of assessment
practices in a rather straightforward, bivariate manner. The world,

however, is more complex than that. Sometimes when we extract one or two factors from the larger number of possible relationships, we distort their effect, either enlarging or lessening their impact. When one variable is controlled for the effects of a third, different—sometimes strikingly different—patterns may emerge. We can now examine the collective impact of various elements on the assessment patterns in San Antonio.

The conventional statistic used to describe the sign and size of the relationship between some independent variable and a policy variable is the correlation coefficient, upon which we have so far relied heavily. This is a useful and simple device for bivariate analysis. But despite its familiarity, it measures only one's confidence that a relationship exists, while we are also concerned with the strength of a relationship. The slope of a regression equation provides an indicator of the strength of a relationship, but the value of a slope is affected by diverse units of measurement, ranging, in the present study, from dollars to ratios.

For a multivariate analysis, the beta weight, also called the coefficient of relative importance, will serve our purposes nicely. The beta weight is similar to, although not identical with, both the partial correlation coefficient and the partial slope in a multiple regression equation. The beta weight measures how many standard deviation units the dependent variable in a multiple regression equation changes with a change of one standard deviation unit in the independent variable, controlling for other independent variables.[11] Like the correlation coefficient, therefore, it is a standardized measure and its values are not distorted by comparing one variable measured in dollars to another measured in percentages. Its principal advantage for our purposes is that it will enable us to ascertain the relative importance of particular variables while controlling for other variables.

It is plainly impractical to include the universe of independent variables available to us. We follow Stouffer's dictum that "exploratory research is of necessity fumbling, but . . . the waste motion can be reduced by the self-denying ordinance of deliberately limiting ourselves to a few variables at a time."[12] It is thus appropriate to make some selections about the utility of various explanations and select key indicators for those explanations identified. The selection, however, must be conceptually defensible, enabling us to include a select range of potentially crucial indicators.

The variables included in the multivariate analysis are representative of the four dominant hypotheses about service distribution: the socioeconomic hypothesis (median family income, percentage poverty families); ethnicity (percentage minority, percentage Spanish heritage); political power (ratio of political elites to population); and ecological (age and density of neighborhood). Each of these is relatively distinct conceptually. Although percentage minority and percentage Spanish heritage are obviously correlated with one another, one measures the presence of Chicanos alone and the other the dominance of minorities regardless of ethnic characteristic. Median family income and percentage poverty families are obviously inversely related, but the relationship is not so close as to make one merely a surrogate for the other. The measure of political power, however, is sui generis. There is no other measure of political power which is so suitably interval in character, and none for which this could not serve as a surrogate. The age and density of a neighborhood represent two crucial ecological attributes which are positively and moderately strongly ($r = .58$) correlated.

Collectively, these seven variables explain ("explain" in a statistical sense of the word) somewhat less than half of the variation in assessment ratios ($R^2 = .447$). Given the imperfection of the measures both of the independent and dependent variables, this is not perhaps an inconsequential amount. What is most useful, though, is an examination of the coefficients of relative importance in Table 3.3. There the contribution of each variable can be sorted out while others are controlled. It is quite clear that the most important single variable is the percent minority in a census tract, and that the larger the percent minority in the tract, the *lower* the assessed valuation in relation to the census derived estimate. In other words, even when other aspects of the census tract are statistically adjusted—its age, density, income levels, political power, and so on—the proportion of minorities still exhibits a powerful, and negative, association with assessment ratios.

An answer to the "why" question is not immediately apparent from the data. Data unfortunately do not speak for themselves, but only through the pages of their interpreters. Some explana-

Table 4.3: Coefficients of Relative Importance (Beta Weights) for Assessment Ratios and Selected Independent Variables

Variable	Beta
Percent minority	−.509
Population density	−.261
Percentage Spanish heritage	.289
Percentage poverty families	−.202
Age of neighborhood	−.132
Ratio of political elites to population	.107
Median family income	−.150

$$R^2 = .447$$
$$F = 11.9$$

tions, though, can be sought in the decisional behavior of the assessment bureaucracy. Engaging in some educated guesswork about the variations in tax assessment, San Antonio's tax assessor told us that "most new subdivisions are close to market value, old stable areas are underassessed, while your so-called slum areas are overassessed." Why the latter? "Well, it's because they just started to depreciate from the day they were built."[13] Our evidence suggests that the third of his hypotheses was in error, but the reaction may nonetheless reflect a perception of heavily minority areas—"they just started to depreciate from the day they were built"—which explains not their *over*assessment, but their *under*-assessment.

Alongside the impact of the minority percentage of the tract, all other variables seem relatively unimportant. It remains clear that the ecological attributes, density and age in particular, are negatively associated with assessment ratios. What is somewhat more interesting is that, once the minority percentage is accounted for, the percentage Spanish heritage becomes a positive predictor of assessment levels. This, in effect, indicates the enormous contribution to explanation made by the Negro component of the minority variable. In San Antonio, blacks are a relatively small, but extremely segregated and highly visible, minority, far more concentrated than Chicanos. It may well be that the assessment bureaucracy reacts in particular to the blackness of the neighborhood, rather than more generally to its minority status. It may thus be simply that black neighborhoods are the ones which "started to deteriorate from the day they were built," at least from the perspective of the tax department.

What we have shown is a rather curious inversion of conventional wisdom. By no stretch of the imagination or the data can the hypothesis that minorities are discriminated against in tax assessment practices be sustained. The evidence is exactly to the contrary. It remains, though, a very real possibility that neighborhoods heavily populated by minorities conjure up particular pictures in the heads of the public bureaucracies, perhaps analogous to the policeman's perception of black neighborhoods as filled with troublemakers. Such perceptions may lead, paradoxically, to lower assessments in such areas than the true values of property could in fact sustain. In any event, we have shown that the reality of distribution of the burdens was more complex than simplistic hypotheses suggested. After looking at the burdens, we can now turn to the benefits.

NOTES

1. Subcommittee on Intergovernmental Relations of the Committee on Government Operations, United States Senate, *Confidence and Concern: Citizens View American Government,* part I (Washington, D.C.: Government Printing Office, 1973), pp. 37-38.

2. Advisory Commission on Intergovernmental Relations, *The Role of the States in Strengthening the Property Tax* (Washington, D.C.: Government Printing Office, 1963), pp. 3, 4.

3. Brief for plaintiff, by Joel S. Berke, *Rodriguez v. San Antonio Independent School District,* October 1, 1971.

4. The article cited by the Court is Note, "A Statistical Analysis of the School Finance Decisions: On Winning Battles and Losing Wars," *Yale Law Journal,* 81 (1972), pp. 1303-1341.

5. Mark G. Yudof, "The Property Tax in Texas under State and Federal Law," *Texas Law Review,* 51 (1973), p. 887.

6. Author's interview with San Antonio Tax Assessor-Collector, July 13, 1972.

7. See the evidence cited in Yudof, *op. cit.,* p. 892.

8. Advisory Commission on Intergovernmental Relations, *State and Local Finances: Significant Features, 1966-1969* (Washington, D.C.: Government Printing Office, 1970), pp. 3-4.

9. See their two articles, "Assessment-Sales Ratios under the Boston Property Tax," *National Tax Journal,* 18 (1965), pp. 36-40; and "Assessment-Sales Ratios under the Boston Property Tax," *Assessor's Journal,* 4 (1969), pp. 13-29.

10. For a more general assessment of the impact of taxation on the distribution of income, and one finding that the impact is minimal, see Joseph A. Pechman and Benjamin A. Okner, *Who Bears the Tax Burden?* (Washington, D.C.: Brookings Institution, 1974).

11. For a description of the beta weight, see Hubert Blalock, *Social Statistics* (New York: McGraw-Hill, 1960), pp. 344-346. Applications of the beta weight to data on urban policy outputs can be found in Alan Campbell and Seymour Sacks, *Metropolitan America* (New York: Free Press, 1967); and Edmund P. Fowler and Robert L. Lineberry, "The Comparative Analysis of Urban Policy: Canada and the United States," in Harlan Hahn, ed., *People and Politics in Urban Society* (Beverly Hills, Sage, 1972), pp. 345-368.

12. Samuel A. Stouffer, *Social Research to Test Ideas* (New York: Free Press, 1962), p. 297.

13. Author's interview with San Antonio Tax Assessor-Collector, July 13, 1972.

Chapter 5

WHO GETS THE BENEFITS?

Politics: Who Gets What How.

H. Lasswell

Service delivery problems are the most fundamental urban problems. Because urban services are personal, direct and locality-specific, they are—in terms of both delivery and citizen needs—highly divisible.

D. Yates

Urban public services are basically of two types. Fixed, immobile facilities like fire stations, parks, and libraries are capital intensive and sited (in a city large enough to have more than one of each) at various points around the community. Their utilization very often depends upon a mixture of availability and citizen motivation. Other services are delivered directly to the citizen and are typically mobile and labor-intensive. Policing and garbage collection are the best examples. Frequently, the consumption of these services is virtually automatic. Police patrols do not depend upon citizen requests, but deliver services regardless of citizen demands for services. To be sure, almost all services entail a certain level of citizen initiative. People may call the police to report a crime or request service; people may request special garbage pickups.

In this chapter, we utilize examples of both types of services, and include a range from immobile, capital-intensive services, like parks and libraries, to mobile and labor-intensive ones like police. For both, we continue our emphasis on two principal questions. The first is *whether* the service is equally or unequally distributed. The second, assuming that the answer to the first is that the service is not equally allocated, raises the problem of *why*. There are four hypotheses which we set out in Chapter 3 which can be tested with the data we have in hand. The uncerclass hypothesis, a composite of three particular hypotheses about race, class and political power, suggested that inequalities were a function of one or more sociopolitical characteristics of neighborhoods. Alternatively, we suggested that certain attributes of neighborhoods themselves, apart from their demographic composition, were related to the service largesse they enjoyed. Though this ecological hypothesis potentially has numerous components, we utilize the age and density of neighborhoods to stand for their ecological attributes. We are now ready to provide some evidence on these questions and hypotheses.

Fixed Service Facilities: Equal or Unequal?

The fixed, capital-intensive services we investigate include fire protection, libraries, parks, water, and sewers. About each, we inquire into their distribution from neighborhood to neighborhood in San Antonio.

Inequality is another way of saying dissimilarity. There are a number of ways to identify dissimilar or unequal patterns across units of analysis.[1] Perhaps the easiest is simply examining the *range* of extreme values of service variables, such as linear distance to a library among various census tracts or their differential rates of public water supply. Table 5.1 does indicate the extreme values in the ranges of various dependent variables. Sometimes, these differences are large indeed. Citizens in one census tract live as close as a quarter of a mile from their fire house, while citizens in another live two and three-quarters of a mile away; citizens in one census tract have access to a park with 19.62 acres of developed area per 1000 population, while others live closest to a vest-pocket area of a tenth of an acre per 1000. These are wide deviations indeed. Yet ranges as a measure suffer the disability of saying both too much and too little. They may magnify the extreme differences without telling how typical or atypical the extreme case in fact may be. That human heights have ranged (according to one's Guinness) from 8'11" to 23.2", glosses over the fact that the overwhelming majority of people one meets on the street range from, say, five to six and a half feet.

Consequently, Table 5.1 also provides a more useful statistic for showing dissimilarity. The coefficient of variation (CV) is statistically no more than the standard deviation divided by the mean, but its utility lies in comparing the relative dispersions or dissimilarities of different service measures.[2] Because it is a relative measure, having no fixed range (say, every value falling neatly between zero and one), it can only be used to compare one CV to

Table 5.1: Means, Standard Deviations, Range, and Coefficients of Variability for Measures of Service Distribution

Service Measure	$\bar{X}$	s	Coefficient of Variability (CV)	Ranges Max.	Min.
Fire					
Fire distance	1.04	.52	.50	2.75	.25
Libraries					
Library distance	1.73	.96	.55	4.25	.25
Library volumes per capita	1.09	2.06	1.88	9.03	.34
Library personnel per 1000 population	.72	1.32	1.83	5.84	.17
New books per 1000 population	.45	1.04	2.32	4.33	−.08
Library expenditures per capita	1.85	4.09	2.21	17.66	.48
Parks					
Park distance	.99	.64	.66	3.40	.15
Park development	4.51	5.43	1.21	19.62	.10
General evaluation	5.68	1.93	.34	8.00	1.00
Pool distance	4.57	3.13	.68	9.00	1.00
Pool quality	7.50	.64	.09	9.00	6.00
Playground quality	4.85	2.90	.54	8.00	1.00
Sportsfield quality	4.97	2.86	.58	9.00	1.00
General evaluation	5.68	1.93	.33	9.00	1.00
Water					
Percentage housing units with public water	99.18	1.73	.02	100.00	89.90
Sewers					
Percentage housing units with public sewers	95.82	7.57	.08	100.00	52.9

another. It is somewhat unusual, however, to find such coefficients larger than 1.00, yet there are a number of that magnitude among these measures of services. By way of providing a baseline, it might be useful to compare these CV's with those for the independent variables listed in an earlier chapter, specifically in Table 3.5. For the most part, the indices of neighborhood attributes (race, class, and ecological composition) were less than 1.00, although those for political power were far higher, indicating, as we remarked at the time, that the political power scores of various neighborhoods were very dissimilar. Among these dependent variables, *there is just about as much dissimilarity among neighborhoods in the amount of services they receive as among their socioeconomic, racial, and ecological attributes.*

The assumption is warranted that urban public services are not equally distributed in San Antonio. Thus far in our analysis—though we have just scratched the surface of the real issues—we can offer some tentative support to those commentators recounted in Chapter 1 who held that the urban public sector is not all share and share alike. Such a discovery tells us something, but surprisingly little. Services may be distributed unequally for a variety of reasons, many of them perfectly defensible, some of them utterly malevolent and possibly unconstitutional.

Parks and Recreation

One can construct an argument on either side of the hypothesis that park and recreational services discriminate against the urban underclass. A variety of evidence derived from the judicial resolution of conflicts over urban services suggests that parks can frequently be shown to be "direct" in their relationship to socioeconomic traits like income, education, and occupation, as well as negatively related to the proportion minority in a neighborhood. In cases involving Prattville, Alabama, and San Francisco, California, parks favored the already favored. In Prattville, the ratio of white to black population was one to five, while the relation of white to black park acreage was fifteen to one. In San Francisco, the pattern of recreational facilities available to the Chinatown population was substantially below the size and quality of those available to the rest of the city's population.[3] On the other hand, not all studies of park services draw the same inferences about discrimination against the urban underclass. Capital spending on recreational facilities in the City of Brotherly Love shows a relatively equivalent pattern from rich to poor neighborhoods.[4] In Detroit, to the degree that park services are biased at all, they tend to favor the poorer neighborhoods.[5]

Here, we should be able to settle, at least for the instant case of San Antonio, the debate about the distribution of park services. We focus not only on the dimension of park quality, measured by such factors as evaluator scores, developed acreage, and the like, but also on the problem of accessibility. Overall, the city of San Antonio maintains a relatively effective and creative parks and recreational program, especially considering the relative poverty of the city's public sector. The city owned in 1972 about 4400 acres of parkland in 82 public parks, which works out to about 6.4 acres of parks per 1000 persons. This is somewhat below the national standard advocated by the National Recreation Association of 10 acres per 1000 persons. But what the city lacks in quantity, it tries to make up for in quality. The keystone of the San Antonio park system is Brackenridge Park, a gift of a wealthy benefactor in 1899, a park which parallels the richness and diversity of Chicago's Lincoln or Mexico City's Chapultapec. Brackenridge's zoo is among the nation's finest, the park's landscaping among the most extensive.

We are not, however, really concerned with the full range of park and recreational sites in San Antonio. We eliminated from our attention span such things as golf courses, burial plots, cultural facilities, and the sui generis Brackenridge, and focused exclusively on those thirty-four sites designated by the parks department as either community or neighborhood parks. For these we were able to secure a battery of information, not only about their distance and accessibility, but, perhaps more importantly, about their quality and utilization. Fortunately for our purposes, in the summer of 1969, the Texas Department of Parks and Wildlife conducted an inventory of every park in Texas, counting and evaluating every park, its equipment and facilities. The resultant information for San Antonio parks provides a very considerable range of data.

Much of it includes quality measures, however subjective they may be, on the accessibility, overall condition, and utilization of parks in the city.

The battery of park indicators is summarized in Table 5.2, which shows seven measures of park accessibility and quality. Throughout the analysis of parks (and other services reported in this chapter as well), the unit of analysis is the census tract. The 113 tracts which are wholly within the borders of San Antonio (and excepting tracts which are coterminous with military bases) are scored with the attributes of their most proximate parks. Of course, people may not use the nearest park, preferring to drive somewhere else to enjoy recreational facilities. If people take this option, something may be said about their perception of the nearer park.

Two of the measures (park and pool distance) are straight-forward "as the crow flies" mileage estimates from the tract as a whole to the park serving it. Not blessed with the energy to count mileage from every household, we took six spots in each tract, scattered about the tract randomly, and calculated their average linear distance to the applicable park. The measure of park development is a per capita measure, based upon the ratio of the developed acreage in the park to the summed total of the population of the census tracts most proximate to it. It is necessary to weight such measures as size of park to a population base, for to do otherwise would mean that two parks of equal size might serve very different numbers of households.

The four quality indices are derived from the elaborate (and by all appearances, very exhaustive) study of parks in the state of Texas by its parks and wildlife department. They rely upon enumerator estimates of park quality. There is nothing sacrosanct about these ratings and it is only fair to observe that most San Antonio parks fell into the five–eight range of good to very good. However subjective, they will at least give us a constant baseline of comparison, whereby we may examine as dependent variables an estimate of the quality of the most proximate park to each census tract in San Antonio, and relate that quality ranking in turn to the socioeconomic, racial, political, and ecological attributes of the tract.

Table 5.2: Service Measures: Parks and Recreation

Indicator	Definition and Measurement
1. Park distance	Mean distance of six random points in census tract to nearest neighborhood or community park
2. Pool distance	Mean distance of random points in census tract to nearest public pool
3. Park development	Number of developed park acres per 10,000 persons served, i.e., ratio of developed acreage to population of census tracts most proximate to park
4. Park general evaluation	Enumerator overall assessment scored from 1-9 of park quality
5. Playground quality	Enumerator assessment scored from 1-9 of quality of playground equipment
6. Sportsfield quality	Enumerator assessment scored from 1-9 of quality of sports facilities
7. Pool quality	Enumerator assessment scored from 1-9 of quality of pool

THE DISTANCE DIMENSION

The utilization of parks, as with other public facilities, may be associated with their accessibility. People are naturally willing to drive for a Sunday at the park to a location somewhere in another quadrant of the city. A Central Park, a Lincoln Park, a Tiergarten, a Chapultapec, or even a Brackenridge cannot be duplicated in every neighborhood. The use of neighborhood parks, however, is more likely to be influenced by a distance dimension, especially when the potential users are small children or teenagers of the pre-automobile age cohort. Pearsonian correlation coefficients will show—and they do in Table 5.3—the relationship between these mean distances by tract and the socioeconomic and ecological attributes of the census tracts. There, stronger relationship indicates that more of the attribute (say, higher median family incomes) is associated with greater distances from the nearest public park.

Fortunately, these correlations are not only strong by the usual experience of the social sciences but wholly consistent. None of them drop below a statistical significance level of .05 and most of

Table 5.3: Correlations between Socioeconomic Status, Racial Composition, and Political Power of Neighborhoods and their Mean Distance to the Closest Public Park

Independent Variables	r	Significance
I. Socioeconomic status		
1. Median school years	.489	.01
2. Median family income	.567	.01
3. Percentage poverty families	−.414	.01
4. Median value of power-occupied dwelling units	.632	.01
5. Percentage owner occupancy	.192	.05
6. Percentage white collar	.537	.01
7. Percentage overcrowded housing	−.327	.01
II. Racial composition		
1. Percentage Negro	−.166	.05
2. Percentage Spanish heritage	−.428	.01
3. Percentage minority	−.518	.01
III. Political power		
1. Number of electoral elites*	.162	.05
2. Ratio of electoral elites to population	.154	.05
3. Number of political elites	.274	.01
4. Ratio of political elites to population	.270	.01
IV. Ecological characteristics		
1. Age of neighborhood	−.568	.01
2. Density	−.481	.01

the coefficients (except those for political power measures) are greater than .40. They do not, however, provide much support for the underclass hypothesis in any of its three variants. In fact, they contradict it decisively. The higher the social status of the census tract, the *greater* its distance to the nearest public park. A higher proportion of racial and ethnic minorities in a tract is associated with *closer* distances to public parks. The more political power "resides" in a tract, the *greater* the distance to a park. Clearly, poorer, powerless, and non-Anglo neighborhoods are closer to public parks.

Part of the explanation for this discovery might relate to the exceptionally strong and consistent negative correlations between park distance and the ecological characteristics of neighborhoods. One might expect that in old, densely populated neighborhoods parks were less accessible if for no other reason that park construction has been more an object of public concern in recent years, after such areas may have already been settled and crowded. In fact, though, the older and more densely settled a neighborhood, the greater its proximity to a park. Those who live far away from a public park are more likely to be those upper-middle-income families, typically of Anglo heritage, who live in newer, single-family dwelling-unit neighborhoods. It is those areas where park construction no doubt lags behind population growth, and where residential newcomers cannot expect to find a new park constructed before an area is fully settled. The poor, on the other hand, do live closer to recreational facilities.[6]

THE QUALITY DIMENSION

It is entirely possible, of course, that the urban underclass lives closest to parks, but that the parks to which they are proximate are the worst in the city. It may be scant consolation to learn that one has easy access to a deteriorating, tiny, overcrowded park. Consequently, an emphasis on the accessibility dimension is not alone sufficient and must be combined with attention to the quality dimension. This is not the place to rehash the long debate about the measurement of the "quality" of urban services, if indeed any satisfactory measurement can be found. We intend to settle for what we have, and that represents measures (all identified in Table 5.2) of acreage, together with an evaluation of the park and its facilities by the department of parks and wildlife.

Both space and the patience of the reader preclude an elaborate presentation of the results. But the full battery of independent and dependent variables together demonstrate that qualitative aspects of parks vary widely in San Antonio, but that such attributes are weakly if at all correlated with measures of the underclass and ecological hypothesis. Table 5.4 shows the mean values of poverty and Spanish heritage by the qualitative score of pools in San Antonio. Pools, graded by the Texas parks and wildlife survey, were assigned scores from 1-9, depending on their estimated quality. In San Antonio itself, no pool received a ranking lower than six and only one as high as nine.

The evidence in the table suggests a curvilinear relationship between pool quality and the two independent variables. Both the best and the worst pools tend to serve high proportions of poor and Chicano neighborhoods. On the other hand, these pools are

Table 5.4: Estimates of Pool Quality and Proportion of Families in
           Proximate Census Tracts Which Are Poor and Spanish Heritage

Pool Score	Percentage Poverty Families	Percentage Spanish Heritage
	$\overline{X}$	$\overline{X}$
All (113)	18.48%	50.99%
6 (4)	35.49	97.37
7 (53)	15.17	44.48
8 (51)	18.36	49.57
9 (5)	41.18	95.24

exceptional (only one pool scoring at the low end and serving only
four census tracts and only one pool scoring at the high end and
serving five census tracts). The overwhelming majority of the city's
tracts are served by pools in the seven—eight quality range, and the
mean proportions of poor and Chicano citizens most proximate to
them do not deviate much from the city's average.

Table 5.5 takes another slice of the qualitative pie. It isolates a
political power indicator—the number of council members who
resided in the tract from the inception of the GGL in 1955 until
1972—and shows the corresponding values of various park indicia.
A little imagination and a hard look will seem to show a slight
advantage on some park quality measures (particularly the numbers of developed acres of parkland and pool distance) for the
most politically potent neighborhoods, but the cases are thin
(N=2) there. The lowest quality rankings typically belong somewhere in the middle of the distribution rather than to those areas
with absolutely no councilmanic representation for almost two
decades.

A third and last approach to the data on quality of parks is
illustrated by Table 5.6. It takes the most overarching measure—a
general enumerator evaluation of park quality—and breaks it down
by several neighborhood attributes. The pattern, if there really is
any, is difficult to discern. The picture is a cluttered, confusing,
zig-zag one, where the parks serving the wealthiest and second
wealthiest clusters of tracts score two and seven respectively.

It would be a fair interpretation of the data on the distribution
of park quality in San Antonio to conclude that not all citizens are

Table 5.5: Number of City Council Members from Tract (1955-1972) and Various Measures of Park Quality

						Number of City Council Members					
Park Quality Measure	All N=113	0 N=88	1 N=8	2 N=4	3 N=4	4 N=3	5 N=1	6 N=2	7 N=1	9 N=1	17 N=1
Mean distance to nearest pool (in miles)	1.68	1.67	1.44	.81	1.50	2.37	4.00	1.88	5.00	1.20	.65
General evaluation score	5.68	5.68	5.70	6.25	6.25	3.75	3.67	6.00	7.00	7.00	6.00
Quality evaluation of sports field	4.97	4.86	6.63	4.50	4.25	4.33	2.00	7.00	2.00	9.00	6.00
Number of developed acres per 1000 population	4.51	4.30	7.27	6.47	1.30	.90	.93	1.02	.93	19.62	16.57

Table 5.6: Relationship among Neighborhood Attributes and Overall Quality Evaluation of San Antonio Parks

					Mean Value of Variable by Park Score				
Variable	All tracts	1	2	3	4	5	6	7	8
I. *Socioeconomic status*									
Median family income	$7786	$8600	$9985	$9232	$6841	$5972	$7929	$9425	$6663
Percentage poverty families	18.5%	17.8%	5.6%	9.3%	20.5%	28.7%	19.7%	11.9%	20.4%
II. *Ethnicity*									
Percentage minority	60.4%	53.8%	24.3%	44.2%	57.2%	84.6%	61.7%	44.11%	72.0%
III. *Ecological*									
Age of neighborhood	46.60%	18.5%	9.8%	36.6%	30.9%	56.7%	59.8%	48.9%	44.9%
Density	8.2	6.6	4.3	8.4	3.3	12.4	9.2	7.4	8.2
IV. *Political power*									
Number political elite	2.6	1.0	5.0	4.6	.8	.4	.5	.6	.1
Ratio of political elite to population	.5	.1	1.1	.5	.2	.1	.0	.5	.1

served by equally good parks, but that none of the underclass hypotheses seem to offer much help in explaining irregularities in service quality. If it exists at all, a consistent pattern of discrimination is hard to identify. We presented only a fragment of the evidence to demonstrate the point, but the rest of the iceberg is as confused and muddled in terms of patterning as the tip we illustrated. The underclass hypothesis works quite badly—our findings sharply countermand it—for the proximity aspect of parks. The quality of the city's parks is far from equal, but the underclass explanation will not help us much in explaining the inequalities which do exist.[7]

Fire Protection

The problem of fires is not, to say the least, a widely-researched urban problem. Though cities spend more on fire protection than on any services besides policing and highways, and even though the risk of death by fire approaches that of death by murder, next to nothing is known about fire protection as a policy area. Presumably, policing is a matter of relevance to social science, while fire protection is a purely "technical" matter. Yet, the fire risk is particularly concentrated among the urban underclass. In Chicago, for example, half of the deaths from fires occur among the third of the population which is black, and a fifth among the 10 percent which is aged.

We have observed repeatedly that the measurement of public service delivery is a difficult business, and nowhere is it more difficult than in the case of fire protection. According to one of the few academic students of the fire protection process as public policy, Roger Allbrandt, "the ability to suppress a fire once it commences depends upon the time it takes to get the equipment to the fire after it has been reported (response time), the number of men available to fight the fire, the level of training of the manpower, and the water pressure at the site."[8] Nailing down data on fire services, however, is extremely difficult in general and in San Antonio in particular. Fire departments exhibit the peculiar pattern of being a fixed facility whose utility is virtually nil if it is not highly mobile. While citizens go to hospitals, parks, and libraries, fire departments come to them. Fire departments still make house calls. It is partly because of this mobility that fire protection benefits to neighborhoods are difficult to measure. Counting the fire personnel and facilities available in a particular neighborhood is of little use because fire units can readily "double up" in case of a fire or cover one another when one unit is busy. One useful measure is the mean response time to fires in various neighborhoods, but such data are unavailable in San Antonio[9] and not perfect in any case. Under the best of circumstances response time provides information only on rapidity of service and not its quality. The latter type of information requires extremely subtle and technical judgments which most service analysts are not in a position to make.

Consequently, our measure of fire services must be taken as a very rough approximation of fire protection. We calculated, again for six random points in each census tract, the distance from that point to the nearest fire station. Summing and dividing by six provides a mean score for each of the 113 census tracts on distance to the most proximate fire station. Unfortunately, fire equipment does not travel like the proverbial crow, but on public thoroughfares of varying degrees of congestion. Taking all these limits to the measure into account, however, it may still be useful to see whether there are any patterns of discrimination in the spatial dimension of fire protection. Here is certainly an area where one's egalitarian norms would favor the hope that there is not a high proximity to well-to-do, low-density, and new areas. If equality is to be a useful standard in assessing fire protection, it will be more as a negative test. A fire protection system which was equidistant to every census tract would probably fail virtually every test of need and equity, for certain areas—commercial areas, older neighborhoods, densely settled areas, for example—are almost certain to *need* more fire protection than others.

In any event, Table 5.7 shows the correlations between linear distance and various attributes of census tracts in San Antonio. Again, the higher the correlation, the closer the station. As one might hope, the strongest correlations are between ecological attributes of neighborhoods and their proximity to the fire stations. The older the neighborhood, the stronger the proximity

Table 5.7: Correlations between Socioeconomic Status, Racial Composition, and Political Power of Neighborhoods and their Mean Distance to Fire Station, San Antonio, Texas*

Independent Variables	r	Significance
I. Socioeconomic status		
1. Median school years	.269	.01
2. Median family income	.440	.01
3. Percentage poverty families	−.260	.01
4. Median value of owner occupied dwelling units	.443	.01
5. Percentage owner occupancy	.397	.01
6. Percentage white collar	.313	.01
7. Percentage overcrowded housing	−.018	N.S.
II. Racial composition		
1. Percentage Negro	−.127	N.S.
2. Percentage Spanish heritage	−.208	.01
3. Percentage minority	−.282	.01
III. Political power		
1. Number of electoral elites	.011	N.S.
2. Ratio electoral elites to population	−.027	N.S.
3. Number of political elites	.039	N.S.
4. Ratio of political elites to population	.069	N.S.
IV. Ecological characteristics		
1. Age of neighborhood	−.619	.01
2. Density	−.476	.01

*N.S. means "not significant" at .05 or higher.

measure (−.619); the heavier the density, the more proximate the fire station (−.476). Citizens in census tracts scoring high on median family income clearly are a greater distance from fire houses than those scoring low on the income measure.

It bears repeating that linear distance is not, in legal terminology, "best evidence." It is entirely possible that fire equipment can zip to middle-income neighborhoods with much greater dispatch than to more densely settled neighborhoods, even though they are physically closer to the latter. It is, however, important to note what we did *not* uncover. The data do not suggest that fire protection distances are actually closer to neighborhoods with advantages of power and status. That would be, of course, the unkindest service cut of all.

The Library and the Public

The American public library has had a long and essentially noble purpose. Older indeed than almost all urban public services— Benjamin Franklin helped found the semi-public library, yet police patrols and public schools were not inaugurated until the 1830s—the public library has had a special mission: to bring not only reading but learning to the American experience. The relationship between the public library and the delivery of its services to various clienteles has long been a subject of controversy. In the early nineteenth century, a director of New York's Astor library remarked, the "young fry . . . employ all the hours they are out of school in reading the trashy, as Scott, Cooper, Dickens, Punch, and the 'Illustrated News'."[10] More recently, and more importantly for our purposes, the capacity of the urban library to reach a broad cross-section of its citizens, the poor as well as the affluent, the black as well as the white, has been much debated. Some evidence has accumulated to suggest that the library, its services, personnel, and siting are all biased in favor of middle- and upper-income groups. Among professional librarians, the 1963 publication of the American Library Association's *Access to Public Libraries*,[11] created a mini-firestorm of protest. The "access study" surveyed racial discrimination by public libraries and investigated black-white differences in collection size, siting, circulation, professional staff, and expenditures. Examining cities in both the north and the south, it concluded that

> neighborhoods which are predominantly white have a far greater probability of containing a branch library than those which are highly nonwhite. Furthermore, those branches which do exist in predominantly nonwhite areas are generally among the least adequate of a city's library system. The cumulative effect of few branches and inadequate library resources is a clear pattern of discrimination against nonwhite neighborhoods.[12]

At the time of the study, Washington, D.C., for example, contained a library branch in a fifth of the well-to-do tracts, but no branches in a low income area. In Birmingham, there were twice as many branches and twice as many volumes per capita in rich as poor areas. Despite this evidence, however, the report was not well received (in fact, not officially received at all) by the ALA. The ALA applauded those aspects of the study which identified direct discrimination (such as rules against blacks in libraries, illustrated by the opinions of one southern librarian that libraries should not be integrated because "all Negroes are infected with venereal disease"), but rejected those aspects using census data to examine covert discrimination. Subsequently, however, Lowell Martin produced an even more exacting analysis of urban library services in Chicago, and concluded that the pattern of library facilities, personnel, and expenditures favored more advantaged neighborhoods.[13]

Libraries present a difficult issue in the examination of urban service distributions. Reading itself is a human activity not randomly scattered across the socioeconomic spectrum. That the well-educated read more may be taken as one of the more settled propositions in the social sciences. The relationship is better described, perhaps, as a threshold one than as a direct or linear pattern. Generally, that threshold is a high school education. Over that, people tend generally to read one or more books a year, below that, none.[14] Consequently, there is an equity question involved in the normative distributive standard. If reading is class-related, one should perhaps not insist that all classes be given equal shares of library services, when some might indeed prefer to be doing something else with their time (and their tax dollars). Some frankly contend that the library belongs to the readers, and if most of the readers are middle class, then so be it. Banfield, for example, holds that "the proper business of the public library is with the *serious* reader and—assuming that the library cannot be an effective instrument for educating the lower class—with him alone."[15]

At the same time, it is possible that reasons other than the undereducation of the lower class are at least in part responsible for differential reading habits. Differential access is one potentially critical factor. Perhaps more than any other fixed-site urban public facility, the library's location is related to its utilization. As Berelson concluded, "there is a relationship between the use of the library and the distance separating the user from it . . . the closer people live to a public library, the more they will want to use it. The major part of library registrants and users live within a relatively few blocks of the library building."[16] For the public library, the distance-usage gradient is a steep one. If the library is badly situated, especially from the point of view of the less well-educated consumer, the library will be underutilized.

A host of other factors may be associated with differential reading patterns, other than the distance dimension. The size and nature of collections, the quality and quantity of personnel, the attractiveness and comfort of the facility itself, and other matters may also determine usage. The point is that differential utilization may be seen as a function of one of two—or probably a mixture of the two combined—hypotheses. The first is a "user attribute" hypothesis, which holds that lower socioeconomic status neighborhoods consume less library service simply because less educated households read less. The other is a "service quality attribute" hypothesis, which suggests that variations in the quality of the public facility are independently associated with variations in service consumption. We shall have cause subsequently to examine the distance-quality-consumption nexus, but for now, it is appropriate to examine the initial questions of service variability.

LIBRARIES IN SAN ANTONIO

The relative poverty of the San Antonio public sector shows itself dramatically in library services. The distance San Antonio falls short of national performance standards is, unfortunately, an impressive one. National standards of library service are established by the American Library Association for communities in different size categories. While such professional standards have a way of being set always ahead of mean performance—like the mechanical rabbit at the dog track—the gap between them and the performance of the San Antonio library is too extreme to be a mere function of standards higher than modal performance. The ALA's 1971 standards, which coincide with our accumulated data in 1971, recommend, for a public library system serving a million population, an expenditure of $7.69 million, 154 budgeted professional positions, and new materials acquisition of 125,000 books.[17] The actual service base of the San Antonio system is hard to translate into population terms, in part because the city library serves the entire Bexar County area (a trade-off with the county government which operates food stamps for the city as well as the county). But taking the service radius as the Bexar County population of 850,000, the city spent less than $1.5 million, added only 58,000 books, and had less than 60 professional librarians on the staff.

Nonetheless, the principal issue of service distribution is not the overall quality level of public services in a city, but their differential allocation—or, more exactly, the suspicion of differential allocation—to various neighborhoods. The issue thus becomes whether the main library and its eight branches serving various sectors of the community provide roughly equivalent services and whether those services are correlated with the socioeconomic, racial, political, and ecological traits of various neighborhoods. We have already addressed the question of whether services are differentially allocated to neighborhoods (see Table 5.1 and the accompanying discussion). The coefficients of variation in library service measures are large enough to cast doubt upon the proposition of per capita equality. Whether such distributive patterns can be explained by the four hypotheses relating to neighborhood attributes is the problem we are about to tackle.

THE DISTANCE DIMENSION

We followed the same procedure to measure linear distance from census tracts to their closest public library, whether a branch or the main library, as we did with parks and fire services. Six random spots were identified in a tract and a mileage count was taken, summed and averaged for the tract. This figure then became the distance "score" for the tract. For each of the 113 census tracts, the mean distance to the nearest library becomes a measure of library accessibility. The results of this computation and the correlation coefficients which represent the relative proximity are reported in Table 5.8.

Again, the underclass hypothesis does not fare well. In fact, it is sharply rebutted by the correlations between neighborhood attributes and distance. Indices of socioeconomic advantage, including median family income, educational levels, and housing values, are positively associated with distance, which is to say that people from tracts that are high in those attributes must go further to get to a public library. Indices of ethnicity demonstrate that the more minorities in a neighborhood, the shorter the distance to a library. Political power bears almost no connection to linear distance to libraries and the correlations between power attributes and library access hover near zero. Most impressive, however, are the connections between ecological attributes of census tracts and their accessibility to a library facility. Collectively they constitute the most predictive aspects of neighborhood library accessibility. Particularly crucial is the age of the neighborhood. The age of a

Table 5.8: Correlations between Socioeconomic Status, Racial Composition, and Political Power of Neighborhoods and their Mean Distance to Most Proximate Public Library, San Antonio, Texas*

Independent Variables	r	Significance
I. Socioeconomic status		
1. Median school years	.278	.01
2. Median family income	.354	.01
3. Percentage povery families	−.244	.01
4. Median value of owner occupied dwelling units	.332	.01
5. Percentage owner occupancy	.432	.01
6. Percentage white collar	.257	.01
7. Percentage overcrowded housing	.092	N.S.
II. Racial composition		
1. Percentage Negro	−.058	N.S.
2. Percentage Spanish heritage	−.232	.01
3. Percentage minority	−.258	.01
III. Political power		
1. Number of electoral elites	−.112	N.S.
2. Ratio electoral elites to population	−.180	.05
3. Number of political elites	−.056	N.S.
4. Ratio political elites to population	−.104	N.S.
IV. Ecological characteristics		
1. Age of neighborhood	−.551	.01
2. Density	−.494	.01

*N.S. means not significant at .05 or higher.

neighborhood is strongly and negatively correlated with distance, which is to say that the newer neighborhoods are furthest from a library building.

We are beginning to see a pattern emerge with respect to the distance dimension, and one which merits specification here. *It is the older, near-to-core areas which are consistently most proximate to public facilities.* This has been true for parks, fire protection, and public libraries. Most of the public library branches are of recent vintage (mostly constructed in the 1960s). Naturally enough, they have been built where the readers and potential readers are. There is a rough tendency for the branch libraries to "ring" the central city area, and most are concentrated in relatively dense, settled, and older neighborhoods. Such areas, in San Antonio, as in other southwestern and western cities, are precisely those areas where the mobile middle classes are not principally located. Such persons tend to cluster in "subdivisions" on the periphery of the city, and exhibit higher levels of most measures of socioeconomic status, as well as being disproportionately Anglo. The location of public libraries, in San Antonio, at least, appears to reflect the conjoint influences of ecological patterns of growth and density, coupled with decision-rules about the location of public libraries. Library location projections follow a concentric zone theory of metropolitan growth, and assume that library siting will gradually push outward as urban growth pushes outward. But it is the growth which comes first, not the library. Consequently, the middle-income household, which has just purchased a relatively new home in a peripheral subdivision will find that its accessibility—measured as the crow flies, anyway—to a public library is reduced, not enhanced. It may utilize the library in the same proportions as it did when it was situated in the older, denser neighborhood, but its investments in travel time will increase.

THE QUALITY DIMENSION

Again, however, we must entertain the possibility that ready access is negatively related to the quality of public facilities. As with parks, it is possible that the closest facilities are the oldest, most overcrowded, and worst facilities. That the poor and the non-Anglo population lives closer to the library does not mean that the accessible library is as good as one located in an affluent neighborhood. Because we are here concerned with the attributes of the nine public libraries rather than the distances of 113 census tracts, it is not possible to resort to correlation coefficients, which will not support an analysis of only nine cases. We can, however, use certain quality aspects of the library system to investigate the hypothesis that some groups are particularly ill-served by library services. The utilization of the measures of library services listed in Table 5.9 requires essentially the construction of a service district for each public library. Otherwise, the mere cataloging of attributes such as collection size, staff, acquisitions would tell us nothing except that some libraries had more of some attribute than others. Two branches may have equal-sized collections, but if the number of people served by one is twice the size of the other, they are not really capable of providing per capita equality of service. We therefore assigned each tract to its most proximate library, and the collected population of tracts assigned to library "X" became the denominator for computing measures of service equality such as collection size per capita, circulation per capita, and the like.[18] It goes without saying that these measures of library quality do not capture the full range of possible qualitative

Table 5.9: Service Measures: Public Libraries

Indicator	Definition and Measurement
1. Library distance	Mean distance of six random points in census tract to nearest library
2. Library volumes	Per capita number of library materials, i.e., ratio of library collection size to population of census tracts most proximate to library
3. Library expenditures	Per capita expenditure on staff, collection and operation, i.e., ratio of expenditure to population of census tracts most proximate to library
4. Library personnel	Professional staff at library per 1000 population, i.e., ratio of library staff to population of census tracts most proximate to library
5. New books	Net increase (or decrease) in collection size 1969-1970 per 1000 population, i.e., ratio of new materials to population of census tracts most proximate to library

Table 5.10: Mean Values of Ethnic, Socioeconomic, and Ecological Attributes of Census Tracts (N=113) by Library Volumes Per Capita

Census Tract Attributes	All	.34 (13)	.37 (17)	.50 (35)	.51 (19)	.64 (9)	1.04 (8)	1.56 (9)	9.03 (7)
I. Ethnicity									
A. Percentage Negro	8.45	48.17	4.40	.40	8.52	.48	5.83	4.20	.81
B. Percentage Spanish heritage	50.99	33.52	85.69	49.22	35.85	13.34	72.89	39.42	80.71
C. Percentage minority	60.36	82.63	91.07	50.59	45.03	14.49	80.06	44.62	82.60
II. Socioeconomic status									
A. Median family income	$7786	$5964	$6061	$9134	$7792	$12,074	$5578	$8437	$5149
B. Median school years	9.85	9.79	7.39	10.61	10.52	13.00	8.78	12.02	7.67
III. Ecological									
A. Percentage houses built before 1949	46.60	65.97	37.21	32.46	47.68	9.01	82.56	73.54	89.20
B. Population per acre	8.22	8.90	10.74	7.31	6.19	3.46	12.61	10.08	11.60

dimensions. Professional standards relate not only to matters of collection size and staffing, but also to the quality of collections, the ratio of juvenile to adult materials, the training and experience of the staff, the presence of certain ethnic materials in the ethnic neighborhoods, and a host of other items. For our purposes, though, the measures we have selected will serve nicely for an overall assessment.

For the sake of brevity, we will present only the tip of the library service iceberg. It is really of little matter which of the various measures are presented in tabular form, for all of them take pretty much the same pattern. In Table 5.10 the per capita number of library volumes is shown, and the relationship between that aspect of library quality and the mean scores of various census tract characteristics is identified.[19]

The easiest way of characterizing the overall pattern is to describe it as "uneven." The library weakest in collection size per capita is serving a disproportionately black clientele. But on the other hand, when percentage minority as a whole is considered, any hypothesis about discrimination against minorities must explain away the fact that the worst and next worst on one hand, and the best and third best collection sizes, on the other, are disproportionately minority. Even setting aside the main library from the picture (whose nine volumes per capita may make it a special case), things do not become much more crystallized. The third best branch collection serves the wealthiest clientele, but the second best branch collection serves the poorest. Adding back the main library into the equation gives the poor something of an edge, as it is serving the poorest and least well-educated clientele. On the whole, though, it would be stretching a point to make much of a pattern out of these differences.

Just to add a bit more specificity, and to get at the library service question from a different angle, we can relate—and do in Table 5.11—library attributes to two political power indices. Five indices of library services (distance, collection size, professional personnel, expenditures, and new acquisitions) are related to variations in the ratio of councilmen to population and to variations in the number of electoral elites resident in the tract. Shown are scores of these measures for those tracts exhibiting *no* political representation by these measures versus tracts containing *some*

Table 5.11:   Mean Value of Library Service Indicators and Two Political Power Measures

Library Service Indicator	Ratio of Council Members to Population of Tract		Number of Electoral Elites in Tract	
	0 (N=88)	>0 (N=25)	0 (N=59)	>0 (N=54)
Mean distance to nearest library	1.84	1.34	1.84	1.61
Library volumes per capita	1.12	.98	1.24	.92
Professional staff per 1000 population	.75	.64	.83	.62
Library expenditures per capita	1.92	1.58	2.17	1.49
New books per 1000 population	4.75	2.78	5.47	3.38

representation, whether large or small. To the degree that there are any differences apparent, they generally tend to favor those neighborhoods *without* resident political power. It would probably not be warranted to conclude that avoidance of political power is a successful technique for aggrandizing public services, but library services are certainly not allocated to the great holders of political resources in the San Antonio power structure.

Other measures of the underclass hypothesis would have stated distributional patterns slightly differently—some indicating more direct, others more compensatory relationships, between measures of private-sector advantage and public sector advantage—but there is little to be served by an ad nauseam cataloguing of them. The fact of the matter is really quite simple: On the whole, the quality of library services are very weakly related to attributes of neighborhoods in San Antonio. There is nothing particularly "equal" about the distribution of library services, but our capacity to relate these inequalities to neighborhood attributes is strained indeed. The best description would be to call it a system of "unpatterned inequalities."

Water and Sewers

American standards of creature comforts vary, but universal among them is the essentiality of safe drinking water and sewerage systems. In American cities, these functions are overwhelmingly public sector functions. So universal are they, that even in San Antonio, which probably qualifies as the nation's poorest large city—far poorer than Newark, Chicago, Cincinatti, Washington, or New York—still exhibits a census tract mean of 99 percent of the housing units with public water and 95 percent with public sewers. Yet, as with other public services, there are variations among neighborhoods. Because the presence of public sewers and public water supply constitute a kind of *de minimis* test of service distribution, they should provide one further opportunity to test the underclass hypothesis, as well as the ecological hypothesis. Because the answers to the puzzle are so simple, direct, and unambiguous, however, we need not dally over the answers.

The correlations between various attributes of neighborhoods and the percentage of units covered by water and sewer facilities would not be very useful. The amount of variation among tracts is simply too small to meet the requirements of such methods. Such an analysis would, in any case, have little payoff in testing the underclass hypothesis. There is a much simpler explanation.

The proportion of units covered by public water and sewer systems are associated significantly with only two of our independent variables—age of neighborhood and its density. None of the underclass measures differentiate between neighborhoods with and without full coverage by public utilities. All the neighborhoods without full coverage are in the periphery of the city. They are, for the most part, areas which are in the process of transformation from rural to urban, and which score low on the density measure and high on the newness measure. It is not, in San Antonio, the old, core city neighborhoods which get less public utility connections, but newer and sparsely settled areas.

To be sure, the mere presence or absence of sewer and water facilities does not really depict the entire range of possible issues of service quality. In the most notable service equalization lawsuit, "Hawkins v. Shaw,"[20] at issue was not only the absence of such facilities in many minority areas (99 percent of the whites had sewers while only 80 percent of black families did), but also variation in the quality of those facilities. Most of the white community, for example, had 6″ water mains, while most of the black areas of the community had 1¼ or 2″ mains. The size of mains is far more important than merely determining the speed with which one's bathwater runs. Most crucially, it determines the capacity with which fires can be fought, and in turn, the cost of fire insurance. Although we have collected no systematic evidence on the size of water mains, in San Antonio, as elsewhere, there is a strong relationship between age of neighborhood and its water main size. In the Model Cities neighborhoods—which are roughly coterminous with a good deal of the concentrated poverty in San Antonio—water mains tend to be 2″ and a major Model Cities project was to upgrade much of this system with 6″ pipes. Newer neighborhoods, by and large, have bigger pipes, typically 6″ systems. These conditions are probably less a product of purposeful discrimination than of the state of technology and prevailing practice at the time of construction. Water main size is inadequate in many poorer neighborhoods not simply because they are poor, but rather because they are old.

The Consumption-Distance-Quality Nexus

People use or consume a public service for a variety of reasons. Broadly speaking, it is possible to suggest two broad approaches to the question of service consumption. One, the "consumer attribute hypothesis," begins with the attributes of the potential consumer—his or her race, class, taste, or whatever—and assumes that consumption is related to the characteristics of the consumer. Groups do differ in their consumption of public services. There is apparently a very wide gap, for example, between blacks and whites in their consumption patterns. [21] On the other hand consumption habits may be explained by the "service attribute hypothesis," which relates the consumption patterns to the various elements of the service itself. Attractive, effective, and accessible services will be more likely utilized than shoddy public services. Obviously, these assumptions are not incompatible, and the truth is almost certainly that people consume services both because of their own attributes and the attributes of services themselves. If services were utterly nondiscriminatory in their distribution, then the consumer attribute hypothesis would no doubt explain almost all variation in service consumption. Then taste or need alone would determine utilization. People who like to read books would use the library; people who enjoy tennis would use the parks; people who suffer crime victimization would call the police.

There may be, however, distributional or qualitative differences in the delivery of public services which serve to distort the operation of the consumer preference process. The public sector, unlike the private sector, is monopolistic. Interestingly, the virtues of a monopolistic public sector have been disputed in recent years,[22] and the schemes of educational vouchers has not only been touted, but actually inaugurated in at least one school district (Alum Rock, California). Still, the monopolistic pattern pervades urban public services. Monopolies, of course, behave like monopolies and can produce a powerful market distortion effect unheard of in a system of unfettered competition.

The monopolistic character of urban public services means that the attributes of services themselves, as distinct from merely the needs or tastes of consumers, may significantly influence consumption. Two such attributes of services are proximity and quality. People differentially utilize public services, from libraries[23] to health care, in part as a function of proximity. Goving and Coe surveyed the utilization of medical facilities by the poor and concluded that "distance and the related costs of transportation were frequently specified as impediments to proper health care. .... Utilization of facilities becomes, in part, a matter of the ecological distribution of services. Distance most clearly affects the use of medical facilities by Negroes. There is an almost perfect inverse correlation (-.92) between distance and use of public hospitals by the Negro sample population."[24] Quality determinations also influence utilization. As with the private sector, people consume what appeals to them. Where consumption is discretionary, as it is with parks and libraries, quality may be a major determinant of use. Generally, we may characterize these interactions as the consumption-distance-quality nexus.

Our data permit examination of the intercorrelations among distance, quality, and consumption in two areas, libraries and parks. The use of both services is highly discretionary and the "need" for them is not (unlike, say, fire or police services) random or episodic. The data in Table 5.12 bear on this issue by showing the rank-order correlations[25] among use (circulation), quality (here indicated by collection size), and distance (measured by the

Table 5.12: Rank-Order Correlations between Utilization, Distance, and Quality of Public Libraries

	Utilization	Distance	Volumes per capita
Utilization	1.00	.59	.64
Distance		1.00	-.30
Volumes per capita			1.00

mean distances from the census tracts to their most proximate library). Excluding the main library as a special case,[26] these correlations depict the nexus among use, proximity, and quality. Clearly, both distance and quality are positively and strongly related to consumption habits. These patterns suggest that, in San Antonio at least, service attributes are major determinants of utilization of public services. They are, incidentally, far stronger than the correlation between attributes of census tract residents and library utilization. The correlation between median school years (the most useful "demand" measure) and library circulation at the most proximate branch is quite low (.058). In the monopolistic public services market, attributes of public facilities, particularly their accessibility and their quality, are crucial correlates of their utilization.

A similar story is told in the domain of recreational services. The parks and wildlife department survey from which we draw much of our data on park quality also included estimates of facility utilization. The relationships between various quality measures and these utilization estimates are generally strong. The correlation between quality rankings of sports facilities and the usage estimates is .738, between pool quality rankings and usage estimated, .140, and between playground quality and usage, .813. Distance, however, plays a less dominant role in the consumption of recreational facilities. The sheer number of park facilities compared with the relative paucity of libraries (thirty-four community and neighborhood parks versus nine libraries) undoubtedly plays a role. There are simply different economies of scale in the production of park facilities and public libraries. A library with six or eight hundred volumes is not much of a library and will attract very few browsers. But a neighborhood park with even minimal facilities can play a major role in a balanced recreational system.

On the whole, though, there is something to be said in behalf of the "service attribute" explanation for consumption patterns. In the urban public sector, a citizen-consumer will utilize public services not only in accord with his or her own tastes and needs, but also as a function of the quality and accessibility of those facilities available. This suggests again the importance of the analysis of the distribution of public services as a crucial subject or urban political inquiry.

The Location of Fixed-Site Service Facilities: A Multivariate Analysis

Fixed-site public services have one thing in common—citizens must come to them to enjoy the advantages they dispense. (The major exception, of course, is the fire department.) There is a very small number of service facilities which must serve a very large number of citizen-consumers. Their spatial placement thus becomes, as we have just emphasized, a crucial variable affecting access and utilization. This placement, in San Antonio at least, is more commonly a function of ecological attributes of neighborhoods than of their socioeconomic characteristics. Our indicators in the service areas of parks, libraries, fire stations, water and sewers, all suggested that the so-called underclass hypothesis was less closely related to variation in service quality than ecological attributes. Nor did the political power of neighborhoods have anything to do with variations in the quality or location of service facilities. The overall distribution of urban service facilities in San Antonio could best be characterized as "unpatterned inequality." Neighborhoods advantaged in the private sector may be advantaged here, but disadvantaged there, by the public sector.

Social scientists who study the urban political economy have yet to develop an adequate theory (empirical or normative) of public facility location. Harvey remarks on the problem:

> It will not be easy to formulate a theory for the location of public activity. In principle, of course, the problem is exactly the same as it is in the private sector—to find a location pattern which is most efficient subject to a set of distributional constraints. . . . But the problem of finding a solution is theoretically obscured by the quasi-monopolistic

structure of public organization and the inability to find any realistic pricing mechanisms. . . . [T]he state of the art in the theory of public facility location has not progressed much beyond the point of relatively simple model articulation.[27]

In addition to the usual reasons, which Harvey suggests, the relative permanence of service facilities coupled with the relative mobility of households is a further complication. Explaining yesterday's decisions with today's need or demand data does not make much sense in a highly mobile urban community.

Indeed, it is not really the need- or demand-related indicators which have figured prominently in our analysis, but rather factors associated with mobility and growth. Our ecological hypothesis suggested that neighborhood attributes related to age and density—themselves associated with growth and decay—best explained the distribution of fixed-site facilities. These indicators carried their explanatory weight even when all of the underclass indicators seemed to contribute nothing to explaining service variations. Ecological factors were particularly important in predicting distance and accessibility to facilities.

It will be useful, therefore, to move beyond the bivariate level of analysis and see whether the dominance of ecological attributes still holds in a multivariate analysis. We have already argued the virtues of a multivariate approach when we assessed tax assessment practices in Chapter 4. It should be noted that multivariate analysis, where variables are set against one another to weigh their contending explanatory power, does not always make for greater simplicity of explanation. It is rather more like the successive multiplication of complexity than like the gradual peeling away of illusions. Nevertheless, if elements which appeared strong in bivariate relationships remain strong, we can be doubly confident of their importance. If elements which appeared weak in bivariate analysis emerge as significant in a multivariate analysis, it may be that their effects were hidden by some covariation with other elements, and it is best to identify their independent contribution. Once variations in ecological aspects are accounted for, it may then be possible to see a more discernible pattern of service discrimination toward the urban underclass.

In a multivariate analysis, one cannot do everything at once. We have, therefore, zeroed in on a relatively small number of measurable attributes of the neighborhoods, representative of the three underclass hypotheses and the ecological hypothesis. These seven indicators are the same as we examined in Chapter 4 and include measures of socioeconomic status (median family income and percentage poverty families); political power (ratio of political elites to population); ethnicity (percentage minority, percentage Spanish heritage), and ecological attributes (age and density of neighborhood). We can compare their relative contributions to the explanation of service variations using several service measures as dependent variables. Those service indicia most appropriate (appropriate because they represent both linear measurement and a fair range of variation) are fire distance, park distance, library distance, and number of developed park acres.

The stepwise regression analysis procedure works like this: The independent variables are entered into a multiple regression equation by the magnitude of their simple correlation coefficients. As each variable is introduced, the one with the next greatest correlation appears, and so on. One result of this procedure is the production of beta weights (which we described in Chapter 4 and called a "coefficient of relative importance"). A comparison of these coefficients of relative importance is perhaps the soundest test for the influence of one or another neighborhood attribute, because the effects of other variables are accounted for or, in statistical language, "held constant."

Tables 5.13 and 5.14 illustrate what is a generally consistent pattern with these multivariate analyses. The former shows the coefficients of relative importance for one of the accessibility measures, fire distance. In the final equation, median income is the strongest predictor of fire distance, with the two ecological measures next. In Table 5.14, only three of the seven variables seem worth emphasizing. The dominant contribution is by the neighborhood age variable, with minority and poverty indicators also significantly related to the amount of developed park acreage available to a census tract. Never, though, in these or other multivariate analyses, does the political power indicator figure prominently in predicting facility location. The predominant role was more generally played by one or the other ecological measure, with socioeconomic traits variable.

The importance of the ecological traits of neighborhoods, per-

Table 5.13. Coefficients of Relative Importance (Beta Weights) for Fire Distance and Selected Independent Variables

Variable	Beta
Median family income	.647
Age of neighborhood	−.316
Density	−.297
Percentage poverty families	.267
Percentage Spanish heritage	.134
Percentage minority	.117
$R^2 = .474$	

Table 5.14: Coefficients of Relative Importance (Beta Weights) for Developed Park Acreage and Selected Independent Variables

Variable	Beta
Age of neighborhood	.402
Percentage minority	.341
Percentage poverty families	−.299
Median family income	.048
Ratio of political elites to population	.047
Density	−.034
Percentage Spanish heritage	.021
$R^2 = .188$	

sisting even when the effects of other variables are controlled, should not by now be surprising. What does interest the student of urban politics, however, is the utter incapacity to wring much explanatory power out of indicators of political clout. San Antonio is certainly an extreme case of centralization of political power. In very few American cities could a power structure with the potency of the Good Government League be found. The tentacles of such a power clique are long indeed. Acquaintance-ships at banks, civic organizations, and other councils of power make them a formidable force in shaping San Antonio. Their tentacles are not, however, quite long enough to insure their own neighborhoods grander scales of public services. Their role in determining the aggregate level of public policy outputs in San Antonio may well be considerable, because of their influence over tax rates and economic development decisions which in turn affect the size of the public budget. But this does not necessarily grant them power to manipulate the distribution of public services to their own advantage.

Police Services

While parks, libraries, and sewer and water systems are fixed, capital-intensive urban services, policing is a mobile, labor-intensive service. Not only does the police function occupy the plurality share of most city budgets (including San Antonio), the police officer represents the entire political authority structure to citizens. While the most publicized and glamorous aspects of police activities are those related to crime-fighting and detective work, the greatest volume of police work is devoted to the patrol function. Round-the-clock patrol is the oldest urban service and one on which American cities spend an estimated $2 billion per year.

The efficacy of police patrol has recently come under attack. The Police Foundation sponsored an experiment in preventive patrol in Kansas City, Missouri, dividing a portion of the city into reactive, proactive, and control beats. Despite variations in the level of routine patrol activities (from none at all in reactive to heavy in proactive), there were no significant differences in crime rates, feelings of security, victimization survey reports, or even traffic accidents at the end of the year-long experiment.[28] Nonetheless, citizens still *perceive* patrolling to be a major component of the police function. Until alternative policing strategies are devised, we may assume that routine patrol will continue to be a crucial activity of urban police departments.

PATROL AND THE UNDERCLASS HYPOTHESIS

Few issues in urban politics have evoked more discussion than the relations between police and minority groups. Among the many aspects of conventional wisdom is the belief that minority or poor neighborhoods are underpoliced. As with most contentions of conventional wisdom about urban problems, an excellent source is the Kerner commission report, which emphasizes the "relative lack of police personnel for ghetto areas, considering the volume of calls for police."[29] To be sure, underpolicing is not the only source of police-minority group friction. The quality of police contacts, the rapidity of police response, and the unavailability of complaint mechanisms represent other dimensions. The distribution of police manpower is, however, one area where spatial analysis is possible.

As with many other hypotheses derived from conventional wisdom, social science evidence points to a more complex pattern. John Weicher, for example, studied the allocation of police patrols in Chicago and concluded that a curvilinear pattern prevailed with middle class neighborhoods receiving more police patrol than either upper- or lower-class neighborhoods. The Urban Institute compared a lower-income black and an upper-income white neighborhood and found no significant differences in the levels of police patrol. Kenneth Mladenka studied manpower allocations in Houston and concluded that the poorest neighborhoods received the most police manpower.[30] Yet when the Urban Observatories' survey of citizen attitudes in ten cities asked people to compare the effectiveness of "crime fighting" in their neighborhood with the rest of the city, 28 percent of black, compared with only 7 percent of white, respondents said it was "not as good." The beliefs of people cannot be addressed with our data; the distribution of police manpower, however, can be.

METHODOLOGY

The City of San Antonio is divided into nine police sectors, and each sector is further subdivided into 10-12 districts and each of those still further divided into subdistricts. None of these correspond exactly to census tracts, but there is sufficient overlap to match the tracts to their appropriate sectors. Because these matches are imperfect, we cannot confidently attempt to utilize census tract data. Instead, we use only a single independent variable—percentage minority—rather than the battery of indicators we have used earlier. Because tracts with high or low minority percentages are clustered about the city, we can still make generalizations about the spatial allocation of police manpower by sector.

Calculating the assignments of each officer by sector over a single year would be a heroic task. Hundreds of officers, covering four time shifts and nine sectors over 365 days a year produce a mountain of data, one much too large to be climbed. We therefore economized by a straightforward sampling process. The Uniform Patrol Division is divided into four "patrols," lettered A-D. Each works at a single time shift (say, the 7 a.m.–3 p.m. shift) for two months and is then moved en masse to a new time shift (say, 3 p.m.–11 p.m.). Even following a single shift over 365 days by sector would be a massive data gathering venture. We therefore selected 25 random dates during 1974 (drawn from a table of random numbers) and located *each officer each day by sector.* Multiplied properly, the 25-day sample can be generalized to a 365-day year. Because departmental policy requires officers to remain in their sector during their workday, we have a measure of the allocation of patrol services—we call it "man-units"—among the sectors of the city.

As with others of our measures, *raw data* are not very useful by themselves. Clustering massive manpower in a sector with low population, low crime rates, and few calls for service is not the same as clustering massive police manpower in a sector with a large population, high crime rates and numerous calls for service. What is required is a set of denominators, and so we construct three measures of police manpower:

1. police resources as a function of *population,* i.e., police man-units per capita;

2. police resources as a function of *need,* i.e., police man-units per reported crime rates; and

3. police resources as a function of citizen *demand,* i.e., police man-units per number of radioed citizen calls for service (because most calls do *not* involve crime reports, this measure is quite distinct from 2).

Each of these gives us a slightly different angle on the allocation of police resources. Police departments may vary in their maximization of any of the decisional standards. One department may prefer to deploy resources roughly equally among neighborhoods, and thus try to equalize the allocation of manpower per capita. Another department may believe in concentrating resources in high-crime neighborhoods and thus distribute services *very unequally* in per capita terms. Still another may emphasize equalizing the workload of officers and distribute manpower according to the demand, indicated by calls for services, on police time. None of these is inherently more defensible than another, but the various objectives cannot be maximized simultaneously. Most police departments will attempt to work out rough compromises among the contending standards.

MANPOWER ALLOCATION

Table 5.15 takes a very simple classification of sectors, ranked from high to low on their percentage minority, and shows the distribution of police man-units. Three of the nine sectors are 75-100 percent minority; two are 50-74 percent; three are 25-49 percent; and one other sector is predominantly (89 percent) Anglo. These data suggest that there is very little variation among sectors in their allocated levels of police manpower. Regardless of whether allocation is measured by per capita, need, or demand measures, the differences are not great.

There is some clustering of police resources in relationship to crime in the high-minority neighborhoods, but these sectors also include a particular sector with the highest crime rate. In per capita terms, there is "overpolicing" of neighborhoods in the 50-74 percentage range, but those neighborhoods include the downtown area.

To the degree that there are variations in police patrol practices—and the pattern of equivalence seems more apparent than the deviations—they seem to be associated with "overpolicing" downtown and one very high-crime neighborhood, which is also high on percentage minority.

Obviously, there is much to which these data do not speak. They say nothing about the *quality* of police protection, as opposed to its quantity, nothing about the *conduct* of police officers, and nothing about the *efficacy* of police patrol. They deal only with the problem of the spatial distribution of police manpower.

Table 5.15: Distribution of Sampled Police Man-Units, by Percentage Minority

Percentage Minority	Man-Units per Sector	Man-Units per Capita	Man-Units per Sector/Number Major Crimes	Man-Units per Sector/Number Radio Calls
100%–75%	7873.70	.09	1.80	.20
74%–50%	6990.48	.17	1.13	.17
49%–25%	6894.72	.08	1.17	.18
24%–0%	7359.84	.08	1.31	.23

Conclusion

There are numerous explanations for the distribution of urban services. We have focused on a test of several of these explanations. Of the two major types of urban services, fixed and mobile, we found considerable variation in the former. Parks and libraries tend to be distributed in a fashion which might be characterized as "unpatterned inequalities." The underclass explanation, however—neither its racial, nor its power, nor its socioeconomic components—did not offer much explanation. To the degree that an explanation for these inequalities can be found in the attributes of neighborhoods, the ecological traits seem most important.

On the other hand, the mobile service we investigated, police patrol, offered less variation among neighborhoods. The city of San Antonio provides remarkably equivalent allocations of manpower from sector to sector. This equivalence persists whether one uses population, crime rates, or calls for service as a denominator. There is simply not much inequality there to be explained.

NOTES

1. The most useful discussion of the subject is Hayward R. Alker and Bruce M. Russett, "On Measuring Inequality," *Behavioral Science,* 9 (July, 1964), pp. 207-218.

2. A simple and direct discussion of the use of the CV is found in Christopher Jencks, et al., *Inequality* (New York: Basic Books, 1972), pp. 352-353.

3. On the Prattville case, see *Hadnott v. City of Prattville,* 309 F. Supp. 967 (M.D. Ala. 1970). On the San Francisco situation, see Memorandum of Plaintiffs, *Woo v. Alioto,* civ. no. 52100 (N.D. Calif., November 16, 1970).

4. David W. Lyon, "Capital Spending and the Neighborhoods of Philadelphia," *Federal Reserve Bank Business Review of Philadelphia,* (May, 1970), pp. 16-26.

5. Steven D. Gold, "The Distribution of Urban Government Services in Theory and Practice: The Case of Recreation in Detroit," *Public Finance Quarterly,* 2 (January, 1974), pp. 107-130.

6. Though we do not present the data, pretty much the same is true of distance to the nearest pool. Neither the direction nor size of the correlations between pool distance and neighborhood traits varies much from those reported in Table 5.3.

7. This is generally consistent with the evidence on Detroit in Gold, *op. cit.*

8. Roger Allbrandt, "Efficiency in the Provision of Fire Services," *Public Choice,* 16 (Fall, 1973), p. 6.

9. The department records times of calls and times when the units on call officially report a fire to be extinguished, but unless one assumes that all fires take equally long to put out, this will not be a very useful measure.

10. Quoted in Edward C. Banfield, "Some Alternatives for the Public Library," in Banfield, ed., *Urban Government* (New York: Free Press, 2nd edition, 1969), p. 647.

11. American Library Association, *Access to Public Libraries* (Chicago: American Library Association, 1963).

12. *Ibid.,* p. 57.

13. Lowell Martin, *Library Response to Urban Change: A Study of the Chicago Public Library* (Chicago: American Library Association, 1969).

14. *Ibid.,* p. 6.

15. Banfield, *op. cit.,* p. 651.

16. Bernard Berelson, *The Library's Public* (New York: Columbia Univ. Press, 1949), pp. 43-44.

17. American Library Association, *Public Libraries Newsletter,* 10 (October, 1971), pp. 2-3.

18. Because of the service base which is larger than the city proper, we included, in calculating the service measures which are based on a population variable, the populations of incorporated suburbs. Because suburbanites utilize the library, to exclude them from the population base would unnecessarily distort the measurement of such items as circulation, professional personnel, collection size, and new books. Our interest, however, remains with tracts bounded entirely within the city itself.

19. A minor note to clarify interpretation by the particularly careful reader of tables. While there are nine libraries examined here (the main library and eight branches), only eight scores are listed in Table 5.10. As our particularly careful reader of tables has already probably hypothesized, this is because two libraries had an equal number of volumes per capita (.50).

20. *Hawkins v. Town of Shaw,* 437 F. 2nd 1286 (5th Cir., 1971).

21. Herbert Jacob, "Contact with Government Agencies: A Preliminary Analysis of the Distribution of Government Services," *Midwest Journal of Political Science,* 16 (February, 1972), pp. 123-146; Peter K. Eisenger, "The Pattern of Citizen Contacts with Urban Officials," in Harlan Hahn, ed., *People and Politics in Urban Society* (Beverly Hills: Sage, 1972), pp. 43-49.

22. See, e.g., E. S. Savas, "Municipal Monopoly," *Harper's Magazine,* (December, 1971), pp. 55-60.

23. "Libraries are used if the effort required to reach them is within the motivation of potential users." Martin, *op. cit.,* p. 19.

24. John M. Goving and Rodney M. Coe, "Cultural versus Situational Explanations of the Medical Behavior of the Poor," *Social Science Quarterly,* 51 (September, 1970), p. 318. See also, for a brief summary of the evidence, Gary W. Shannon and G. E. Alan Dever, *Health Care Delivery: Spatial Perspectives,* (New York: McGraw-Hill, 1974), pp. 95-102.

25. Here we take a methodological step "down" from Pearsonian correlations to Spearman rank-order correlations. The reason is that the nine libraries and their attributes become the units of analysis, rather than the large "N" of 113 census tracts. An "N" of nine is much too small to support the more rigorous Pearsonian correlations.

26. Actually, it makes very little difference in the relationships reported even if we include the main library.

27. David Harvey, *Social Justice and the City* (Baltimore: Johns Hopkins, 1973), pp. 89-90.

28. George Kelling, et al., *The Kansas City Preventive Patrol Experiment: Summary Report* (Washington, D.C.: The Police Foundation, 1974).

29. National Advisory Commission on Civil Disorders, *Report* (Washington, D.C.: Government Printing Office, 1968), p. 161.

30. John C. Weicher, "The Allocation of Police Protection by Income Class," *Urban Studies,* 8 (October, 1971), pp. 207-220; Peter B. Bloch, *Equality of Distribution of Police Services: A Case Study of Washington, D.C.* (Washington, D.C.: The Urban Institute, 1974); and Kenneth Mladenka, "Serving the Public: The Provision of Municipal Goods and Services," unpublished Ph.D. dissertation, Rice University, 1975.

Chapter 6

PUBLIC BUREAUCRACIES, DECISION-RULES, AND
THE DISTRIBUTION OF PUBLIC SERVICES

*Bureaucracy is a circle from which no one can escape. Its hierarchy is
a hierarchy of knowledge. The apex entrusts the lower echelon with
insight into the individualistic while the lower echelon leaves insight
into the universal to the apex, and so each deceives the other.*

Karl Marx

*In every political institution, a power to advance the public happiness
involves a discretion which may be misapplied and abused.*

James Madison

Marx and Madison have both much and little in common. Both
concur in the class interpretation of political life, though they
diverge greatly on its implications for the organization of polities.
Another element of commonality in the two great theorists of
class politics is their relative inattention to public bureaucracies. It
was Lenin who bureaucratized Marx and Wilson who bureaucratized Madison. Not until the era of Wilson and Weber was the
bureaucracy as a social institution given coequal attention with the
economic classes of Marx and Madison. The class and bureaucratic analyses are not, of course, entirely separable, for the
bureaucracy has always been a middle-class institution. Pirenne
remarked of the medieval city that "nowhere, perhaps, was the
spirit of innovation and the practical judgment of the middle class
more highly manifest than in the realm of administration."[1] In the
American city, the bureaucratization of municipal government was
an outgrowth of the turn-of-the-century municipal reform move-

ment, whose social origins were Yankee, Protestant, and definitely middle-class. Today, a dominant thread in the research on bureaucracies is their "middle-class bias." Sjoberg and his associates observe that "sociologists frequently compare lower- and middle-class culture patterns, but they fail to recognize that bureaucratic systems are the key medium through which the middle class maintains its advantaged position vis-á-vis the lower class."[2]

In terms at least of the quantity of urban public decisions, the vast majority of them are made by public bureaucracies, operating, we shall argue, in relative isolation from other centers of power, and with considerable autonomy. It is our contention in this chapter that the urban bureaucracy's decision rules are crucial determinants of the allocation of municipal services. Saying this is not so much stating an hypothesis as it is stating a truism. The preceding two chapters emphasized a sociospatial perspective of urban services. The essence of the hypothesis being tested was that the allocation of public services followed the lines of social power and dominance (indicated by race, class, and power variation in neighborhoods) and the lines of ecological variation among neighborhoods. Describing first the limits of the sociospatial perspective, we shall argue here that the decisional premises of the public bureaucracies offer a kind of "missing link" in the analysis of allocative processes and outcomes in urban government. Because the public bureaucracies represent monopolistic policies of both production and pricing, their decisional principles become of necessity major determinants of the quality and quantity of public goods and services delivered within the urban area.

The Limits to the Sociospatial Perspective

We have spent a good deal of effort so far in examining the sociospatial distribution of urban services. This investigation was premised on the widespread conventional wisdom—given juridical credence in the "Hawkins" case and related cases—that urban services disadvantaged the urban underclass. Were this the case, we should have been able to find relatively consistent patterns of "direct" allocations, i.e., sociospatial areas advantaged by their race, their political power, or their socioeconomic class, receiving bigger and better packages of services. In fact, however, the public Santa Claus operates less like the private Santa Claus than one might expect. The private Santa Claus delivers his largesse in rough correspondence to family incomes. The public Santa Claus is not much more equal in his allocation, but his largess is only weakly related to socioeconomic status. Ecological factors carry us some-what further than considerations of power and status. Age and density of neighborhoods are, on the whole, more closely tied to their service levels than are the class, racial, and political character-istics of their residents.

Generally, social scientists assume that the measure of success in research is a "variance accounted for" model. The larger the percentage of variance explained in the dependent variable— whether it be school achievement, crime rates, or service alloca-tions—the "better."

The present research, however, probably falls into the genre of a "variance discounted" model. A commonplace hypothesis—that the poor, the minorities and the powerless are left holding the leavings of the allocative system—has been analyzed and found wanting. We simply cannot explain much variation in the distribu-tion of urban services in San Antonio with data on the socioeco-nomic traits of neighborhoods. Because this conventional wisdom is so wide-spread (indeed, we spent much of Chapter 1 recounting it), it is wise to do two things at this point. First, we must reiterate the caveats surrounding the present research—the generalization from a single case, the limitations of the methodologies and measurements, the examination of only a single point in time, and so forth. We should also however, indicate that this is not the only analysis of urban services to stand so defiantly against the canons of conventional wisdom.[3]

Even adding in the ecological characteristics will not account for a major proportion of variation in inter-neighborhood ser-vicing. Though ecological variables fare better than the measures derived from the underclass hypothesis, they still do not account for the major portion of the variation in the service indicia. The inability to account for a majority of the variation in the depen-dent variable results, of course, from a multiplicity of factors, including imperfect measurement of included variables, the effects of unmeasured or unmeasurable variables, and inevitable short-comings of the methodological tools. Try though one might, it is not possible to tease these data sufficiently to account for more than a minority of the variation in the service allocations of the city of San Antonio. In terms of the "variance accounted for" model, there must be a missing link to be identified. In our opinion, this missing link is the bureaucratic decision-rule.

Bureaucracies and Urban Policy

THE EMERGENCE OF THE MODERN URBAN PUBLIC BUREAUCRACY

Bureaucracy is hardly a new invention, but it is an institution which is constantly reinvented. The modern urban bureaucracy was reinvented by the municipal reform movement at the turn of the century. In a sense, it was the goal of the reformers to replace one kind of bureaucracy, the political machine, with another, the public bureaucracy as we know it today. The reformers were themselves products of the bureaucracies of the modern corporation and wished to translate its efficiencies to the public sector. enormous proportion, as they saw it, of the resources of municipal government were devoted to "sidepayments." The machines did not pay people to build streets, but paid off people who built streets. This was, according to the reformers, an enormously expensive and inefficient way to run public governments.

Political scientists, including Woodrow Wilson and Frank Good-now, were at the core of this movement. Their leading organ was not the *American Political Science Review,* as it is today, but the *National Municipal Review.* Its appearance in 1912 represented both a catalyst and a codification of the precepts of reformism. The lead article in the third issue was, in fact, devoted to the subject of "Expert City Management." The author touted the virtues of professional city administration and ended with a dis-quisition of the "City of God." This neo-Augustinian approach to urban government was fully characteristic of the dreams of the reformers. Their program entailed a baker's dozen of reform proposals, including those for the depoliticization of city politics (the at-large and nonpartisan election system), the creation of efficiency (budgeting and accounting, governmental consolidation), and professional management (the manager plan, civil service, and advancement by merit).

Urban reformism represents today one of those cases where one generation's reforms are the next generation's problems. The bureaucratic power generated and insulated by the reforms is at the core of neo-reform. Sayre and Kaufman's standard work on New York City is essentially focused on the public bureaucracies, where the dominant behavioral norms were aggrandizement, insulation, bureaucratic imperialism, and self-protection. "Bureaucratic groups," they say, "especially as they mature in their organization and in their self-awareness as cohesive groups, share with all other groups the aspiration to be self-sufficient and autonomous. In fact, bureaucracies appear to present one of the strongest expressions of this general tendency."[4] New York, as it so often is, may be taken as the prototype. Theodore Lowi, describing the differences between the "old machines" and the "new machines," remarks that "cities like New York become well-run but ungovernable. . . . The legacy of reform is the bureaucratic city-state. . . . The New Machines are machines because they are relatively irresponsible structures of power. That is, each agency shapes important policies, yet the leadership of each is relatively self-perpetuating and not readily subject to the controls of higher authority."[5] Savas and Ginsburg suggest that the civil service may be a "meritless system," and hold that reform of service delivery systems is dependent upon new reforms of the old reforms of the bureaucracy.[6]

San Antonio, of course, is not New York, with its massive and influential civil service unions, willing to strike at the drop of a contract. It does exhibit, however, all the traits of the reformed city: the manager plan (it is the second largest city with a manager plan, edged out only by Dallas), nonpartisan elections, at large elections, the civil service system, and all the accoutrements. A general proposition in the study of urban politics is that reformism maximizes the power of bureaucracies. Greenstone and Peterson observe that "political actors in fully reformed cities do not have the incentives to control bureaucracies that they have in machine cities. . . . The expectation in reform cities that administrative practices should be isolated from political influences enables the mayors to hold their supporters' gratitude simply by demonstrating good intentions, even if they fail to exercise effective control. Too effective an iron hand may actually invoke charges of "dictatorship" and "one-man rule."[7]

In San Antonio, there is very little effort on the part of the elected leadership to manipulate the decisional dominance of the public bureaucracies. This is one of those assertions where quantitative data is harder to come by than anecdotal data. The fiscal 1972 budget—the one which in effect finances most of the allocations studied here—summed to $67.7 million. The budget submitted by the public bureaucracy through the city manager amounted to 590 pages; the changes in it authorized by the city council amounted to two pages. Most of the changes in it ("increase street materials account by $100,000 to offset price increases on annual contracts"; "convention center marquee: $100,000"; "two Attorney II's added to Delinquent Tax Office to improve delinquent tax collections: $21,290"; etc.) were relatively insignificant in terms of the great public issues allegedly facing American cities. The largest single item of increase by the council was a 1.79 million dollar across-the-board wage increase for municipal employees to be paid for essentially by the initiation of a household garbage fee.

HOW BUREAUCRACIES BEHAVE: DISCRETION OR DETERMINISM?

A decade ago, one could safely have said that the study of municipal bureaucracies was almost non-existent, except for Sayre and Kaufman's work on New York. Today, the bureaucracy may still be terra incognita, but much less incognita than before. There is manifest evidence of the discretionary power exercised in public bureaucracies, both at the apex and the lower echelon. These bureaucracies have the function of service delivery to some individuals or groups defined as being "in need." They include, of course, welfare bureaucracies, school systems, police officers, probation officers, public health clinicians, and the like. Collectively, Michael Lipsky describes these as the "street level bureaucrats."[8] These bureaucrats share certain defining characteristics.

They deal regularly with citizens, sometimes in an environment of hostility, operate with relative freedom from direct surveyance (precinct captains cannot ride with each patrolman, nor principals always be in the classroom), and are asked to meet client "needs' even where their resources to do so are extremely constrained and the rules by which they operate either ambiguous or so elaborated and detailed as to be unwieldy.

There is, however, a contending perspective on bureaucratic behavior which focuses not on the discretion but on the predictability of bureaucratic decision-making. In this perspective, the remarkable thing about bureaucracies is not how much of their activity is discretionary, but how much of it is deterministic. The concept of "incrementalism" is crucial to this explanation of bureaucratic, especially fiscal, behavior. There is obviously nothing new about the concept and practice of incrementalism. At the turn of the century, the great Irish student of public finance, Charles F. Bastable, wrote that

> fortunately, the question of expenditure in all its forms does not present itself as a single problem. It would be quite hopeless to prepare a budget of outlay for any country without the aid of the material collected during previous experience. *The great mass of expenditure is taken as settled, and it is only the particular changes that have to be anxiously weighed* in order to estimate their probable advantage. This method of treatment simplifies issues very much.[9]

A neater capsule of the theory of incremental decision-making has never been penned. More recently, Wildavsky's analysis of the federal budgetary process, Sharkansky's work on state decision-making, and Crecine's simulations of urban budget-making, all suggest a routinized, highly predictable, and stable-state system of bureaucratic decision-making.[10]

Budgeting in San Antonio works incrementally, much like Crecine's description of Cleveland, Pittsburgh, and Detroit. Urban political systems follow a set of "iron rules" with respect to allocative decisions. The constraint of a balanced budget, the use of past allotments as a baseline for future allocations, the fixed set of priorities (salaries and wages first, maintenance last), and the virtual rubber stamp role of the municipal legislature, all follow consistent patterns in the three cities, and in San Antonio as well. The city manager in San Antonio is the chief administrative officer regarding the budget, rather than the mayor as in some other cities, and he and his department heads dominate the process from the budget's cradle to its grave. Neither the public at large, interest groups in particular, nor even the city council play much role in the allocative process. By the time the budget becomes a public document, most of the decision-rules have been announced, and trade-offs among agencies have already been made. An interest group, therefore, would have to exert influence on the impending decisions at an early stage, or be extremely powerful to force last-minute changes. In the budget year 1971-1972, the only groups able to force alterations at the councilmanic level were upper-status, "civic affairs" groups, including the Symphony Society, the Zoological Society, the Chamber of Commerce, and the Museum Society. In the final analysis, their role in altering decisions at the last stage (less than $100,000 added in a total budget of almost $68 million) was trivial. As for the council's general role, one San Antonio budget officer observed that "the council doesn't interfere."[11] Consequently, the budgetary process represents a kind of life force within the urban political process, operating pretty much according to its own internal logic. This paints a considerably different picture of organizational behavior in city government than does the discretion perspective.

The most obvious way of reconciling these two seemingly divergent explanations of bureaucratic behavior—somewhat akin to the free will versus determinism argument—is to point out that the latter deals with money and its allocation, the former with nonfiscal behavior. This is an explanation we reject. A great deal of nonfiscal allocative behavior is no less routinized and predictable than budgetary processes; the latter is merely more readily susceptible to quantitative measurement and prediction. Actually, a good deal of what passes for discretion is no more than the discretion to adopt one or another routine. At the same time, the budgetary analysis school provides an overdeterministic impression unless it can explain where the allocative rules come from. At the extreme, each perspective contains the seeds of its own contradiction. Bureaucratic behavior in the allocative process cannot be fully explained through the discretionary rubric, because treating every case as sui generis is, almost by definition, nonbureaucratic behavior. Neither can the deterministic model be fully sustained, for it assumes that "forces" or "processes" govern decisions, without inquiring into the decisions which set such forces in motion. We prefer a middle ground. We conceptualize the bureaucratic allocation process as a set of interdependent decision-rules which—though difficult to describe as a collectivity—determine the distribution of municipal services to sociospatial groupings within the community.

THE BUREAUCRATIC DECISION-RULE HYPOTHESIS

Writing in the "Foreword" to Orlando W. Wilson's study of the distribution of police patrol forces in Wichita and San Antonio, published in 1941, the Director of Public Safety of Cleveland, Ohio, remarked that "every police department believes it suffers from insufficient manpower. Every police executive, wishing to meet his complex problem, has been faced with the difficult job of most effectively allocating his personnel. The need for making this distribution according to a plan or system is obvious." [12] With these words, Eliot Ness described the inevitability of bureaucratic decision rules in the allocation of public services. Bureaucratic decision-rules are the minutiae of public administration. They are—to use a phrase from V.O. Key in another context—the "standing decisions" of public authorities about how scarce resources will be allocated. They need not contain any spatial component whatsoever, but they will have often unintended spatial implications. The outcomes of decision rules are the resultant allocations of particular services.

The closer one gets to a bureaucracy, or the further inside it one goes, the more one is overwhelmed by the minutiae of the bureaucratic labyrinth. Dozens of lower echelon employees make myriads of little decisions, cumulating to allocative patterns. Made once, a decision is an exercise of discretion; made twice, it is precedential; made ad infinitum, it is a decision rule for the treatment of classes of cases. Decision-rules result from some rough admixture of professional norms, rules and regulations of superordinate bodies, loose perceptions of both needs and demands, and a search for economizing devices when perceived demands exceed perceived capacity.

In San Antonio, during the year 1971, there were 274,053 incidents in which a police report was filled out. The great majority of these involved a citizen demand for service. Among the 26,500 accidents, the 880 vagrancies, the 6400 prowlers, the 13,700 family disturbances, the 400 assaults, and the rest of the offenses, some allocation of police resources have to be made. Bureaucracies cannot do everything at once. Among the nearly quarter-of-a-million properties on the tax roll of Bexar County, only a handful can actually be assessed and systematically reviewed during a single year. Priorities must be established, for the resources of any bureaucracy are finite and the assessment office unusually so. For the 42,000 tons of asphalt allocated to the gravel and asphalt maintenance of the public works department, someone must decide which potholes get patched. Of the 125,000 new books and materials purchased by the public library, somewhere a decision must be made about the location of those materials by library. How, therefore, do public bureaucracies make allocative decisions where the potential demands for their services exceed the possible scope of the agency's output, and where contending allocative decision principles are available?

There are two remarkable things about decision-rules in the bureaucratic allocations of public services. The first is how low in the bureaucratic hierarchy decision-rules are formulated and implemented. The second is how insulated the rule-makers are from external constraints, an insulation fortified by the relative invisibility of both the rules and the rule-makers. Take, for illustrative purposes, the way police departments allocate their scarce resources. With more than a quarter-million incidents requiring police attention, it is hardly possible that a simplistic egalitarian principle will be of use. Deviations from the attention rule of equal response will be justified—must be justified—by some hierarchy of "importance" or "seriousness." Discretion in responding to police calls is decentralized all the way down to a complaint clerk, operating under a set of decision-rules regarding the seriousness of the complaint. Mladenka describes the process in Houston:

The complaint clerks (both uniformed police officers and female civilian employees) exercise considerable discretion in determining whether a call for assistance enters the dispatch queue. Although a uniformed supervisor (with the rank of police sergeant) is always present in the complaint room, several days of observation revealed no instance in which a clerk's decision not to dispatch a car in response to a request for assistance was overruled. The complaint clerks also determine the nature, and thereby the priority of a call, by checking the appropriate box on the dispatch slip. . . . The clerk's decision to code a call as "see complainant" rather than as a prowler report can have a significant impact upon response time. On occasion, the clerks bypass the queuing system by walking rather than conveying the complaint slip to the dispatch room.

The complaint personnel do not specifically assign a priority to each incoming call. With the exception of those instances where clerks bypass the conveyor belt carrying the complaint message to the dispatch room, dispatch slips are transmitted on a first come first serve basis. The dispatch personnel themselves (all uniformed officers) determine a response priority.[13]

The process of police response works in much the same way in San Antonio. Response hierarchies are, however, neither ad hoc nor complaint specific. There is a routinized hierarchy (where, not surprisingly, "officer in trouble" ranks first, and "family disturbances" quite low) of attention rules. The most important element provided by the lower echelon, therefore, is not really discretion—as is implied by the literature on bureaucratic discretion—but a *definition of the situation.* With extremely fragmentary information, the lowest elements of the bureaucracy (and probably no role could be lower on a police department hierarchy than "female civilian employees") are nonetheless providing, by defining the situation, an agenda-setting role for the entire police department. It would hardly be accurate to say that low-level bureaucratic employees are free to invoke any response they prefer. Street-level bureaucrats can be fired, reprimanded or transferred for creating their own decision-rules. (Witness the case of the New York social worker who assigned his client-recipients to a posh New York hotel and had the department pick up the tab.) Their power rather lies with their capacity to define the situation, thus more or less automatically invoking one or another decision-rule.

THE DECISIONAL PREMISES

When bureaucracies need to make allocational decisions—which is almost all the time—they need rules for doing so. The number of rules which theoretically could be invoked is large indeed. But we suggest that such rules can take five forms, those depending on *demand,* on *professional norms,* on some conception of *need,* on *equality,* or rules which simply react to *"pressure."* Because organizations cannot by definition maximize more than one value at a time, there are inevitable conflicts among these decision premises. Some will be consciously traded off against others; and some will simply be unconsciously followed solely for the convenience of the bureaucracy itself.

The rule of demand is what Frank Levy and his associates have called an "Adam Smith Rule: When a customer makes a 'request,' take care of him in a professional manner; otherwise, leave him alone."[14] Two of the three public services examined by Levy's Oakland project were allocated essentially by Adam Smith rules. Libraries allocated new books largely upon the basis of circulation at branch libraries, and street repairs were allocated primarily on the basis of traffic counts in neighborhoods. Neither of these involved explicit verbalized "requests," as does, for example, calling the fire or police departments. But they both depend upon letting the "market" for public services in various neighborhoods determine the amount of services provided to the neighborhood. The allocation of demand-based public services thus becomes a very straightforward function of *past allocations* (which tend to be rather immutable, as highways, libraries, and so forth are rather permanent) and *very recent consumption patterns.* Demand as a market process described by Levy is not, however, quite the same thing as demand in a more literal sense. People do make individual requests for service to police and fire departments and even to other departments of municipal government. We measured, in Chapter 5, the relationship between police manpower distribution and calls for service. It was rather strikingly similar from sector to sector. Whether consciously or unconsciously, the department in San Antonio responded relatively well to a demand-based measure of service allocation.

There is, of course, nothing inherently discriminatory about delivery of services in response to demand. But when demand itself varies by race or class, services vary by race or class. If better-educated people read more library books, or drive their cars more often, or call the police more frequently, they receive more services. Bureaucratic responsibility for equitable delivery of services is transferred to the marketplace for public services, and some groups play the market more effectively than others. The City Attorney of Los Angeles even advised the city council that it was in violation of the equal protection clause of the fourteenth amendment by its library allocation policies. Like most cities, it had simply allocated books to branches in proportion to use. As the city attorney's report observed:

> The primary reason for existing disparities in the delivery of library services is the use of home book loan circulation as the criterion for allocating library resources. . . . Circulation is a racially neutral criterion for allocation.
>
> Yet that criterion, however neutral on its face, has been unfairly administered in a manner that is discriminatory against minority communities. . . . Resource disparities in libraries serving minority communities have become self-perpetuating. More heavily used libraries attract additional funds and with these additional funds hire more staff, more books and materials, and stay open longer which in turn generates additional usage.[15]

Demand measures, though sometimes defensible, also have allocational implications which are sometimes discriminatory.

Other decisional premises are based upon professional norms and judgments about the "proper" distribution of public services. Numerous professional associations, such as recreation administrators and librarians, draw up national criteria describing the amount of services which should be provided. Like most "goals" or "standards," they are carefully set just a bit ahead of where the average city is likely to be at any particular time. The kind of hierarchies of crime "seriousness" which police departments develop represent professional opinions about the kinds of crimes most in need of immediate attention. Fire departments universally take as an operating norm the location of each fire station so that no area of the city is more than three minutes from a station house. These professional norms serve several functions. First, they give administrators bargaining points when budgets are being considered.

Administrators can either deplore how far the city is behind professionally-set standards or urge the political decision-makers to maintain the city's superior position ahead of other cities. Secondly, they provide rules of thumb for the allocation of sources within the department, and to the neighborhoods.

Need is a very difficult criterion to discuss without getting enmeshed in debates about whose needs are genuine and whose represent merely their self-interest. Yet to dismiss need as a criterion would be to dismiss highly consensual rationales for the distribution of fire, police, recreation, and other services. When police locate units to be closest to high-crime areas, they are responding to need measures. One of the reasons why police allocations diverged from an equitable distribution was because they clustered resources in downtown and in high-crime areas. When fire departments cluster resources in downtown and older neighborhoods, they use need criteria. Libraries are perhaps the best example of a public service which does not respond well to need principles. In San Antonio, as elsewhere, allocation of personnel and resources to high-consumption branches violates a need standard narrowly conceived.

Urban services can also be allocated by the spirit of equal treatment from group to group and neighborhood to neighborhood. It is frankly difficult to say whether such equality as exists is more an accidental or a conscious process of allocation. There is not, as we suggested in Chapter 5, much equality of services in the capital-intensive, fixed-service domain. But there is not much patterned inequality either. We would speculate that equality of treatment is a secondary value of urban administrators, one which might emerge from other decisional premises, but which is not an overarching value in service allocations. If equality premises are indeed reflected, they are probably reflected more in response to the need for political support than on their own terms.

The need for political support leads to the allocation of services on the basis of "pressure." Sometimes squeaky wheels get greased. In a system of at large elections, like San Antonio has, it is not so politically essential for services to be proportionately allocated. Years of dominance by an entrenched power structure reduces the need to respond to pressure. As in most manager cities, council members relay requests for service or complaints to the manager or the appropriate department head. These requests are episodic. In recent years, and since the data for this study were collected, the dominant position of the Good Government League has been emasculated at the ballot box. Political scientists have long hypothesized that competition makes government more responsive to its citizens.[16] If it does, then one might hypothesize that pressures get translated into policy more effectively in competitive local political systems.

None of these decisional premises, however, deny the dominance of the urban bureaucracy in allocative decisions. It is precisely because decision-rules with significant allocative consequences are formulated up and down the bureaucratic hierarchy, that the allocation of public services is sometimes beyond the purview of even the strongest elite. We found no evidence in San Antonio that the political power clustering in sociospatial areas was related to service advantages—a slight tendency, in fact, in the other direction was observed. Too many decision-rules are formulated by too many functionaries to be readily subject to elite control. Indeed, it is not unusual for the lowest operatives to allocate their own time and public resources. Street light installation is pretty much, in San Antonio, a one man show, with a single functionary developing allocative principles (locate a street light on call, provided that the location is at the end of a block and no closer than 300 feet to another light) and adhering to it.

All of this, incidentally, is riddled with implications for the long-standing debate about community power structure. The case of San Antonio is an excellent test of any and all hypotheses about the policy impact of a power structure. Having as it does a monolithic, private-sector elite which exercised enormous influence over elections and victories, San Antonio should give us an idea of the real capacities of a power structure to extend its tentacles beyond the outcomes of elections and into the allocations of public benefits. Frankly, we are more impressed by the limits to the power of the power structure than by its presumed omnipotence. In all probability, the influence of the power structure is superordinate with respect to the tax rate, and the revenue rate strongly determines the aggregate level of expenditures. Concerning the overall budget, the things it purchases, and how they are allocated, the role of the power structure can only be called incidental. The power structure in San Antonio, and elsewhere, is a force to be reckoned with on the "big" decisions. The "little" decisions, however, are made elsewhere, and the cumulative impact of myriads of little decisions is surely as significant as periodic muscle-flexing on the so-called "key" decisions.

Bureaucracies, Democracy, and Public Services

DEMOCRATIC THEORY AND BUREAUCRACIES

Traditional democratic theory has not fared well at the hands of empirical political scientists. The notion of a reasoned electorate making demands by arising on election day to select among contending candidates with contending philosophies has received so much refutation as to make it more a straw man than a descriptive theory. The gist of democratic theory has always been the assumption that citizens wants, needs, tastes, or preferences have a monolithic existence and therefore policy-makers should translate them into public policy. In order to make our argument without having to resolve what it is that citizens are inputting, we merely label this, for sake of convenience, the wants-needs-tastes-preference assumption of democratic theory. Someone else can unravel which of these citizens are supposed to be articulating.

Unfortunately, in urban governance, the democratic model fares perhaps less well than elsewhere. In part, this is because the rate of electoral participation is so low. On the average, in cities over 50,000, only about 32 percent of the eligible electorate turn out to vote. In San Antonio, the proportions are much lower. Only once in the period 1951-1971 did the turnout even approximate that level (1971 with 31 percent participation), and in every other election, it was less than a fifth of the adult population. In American cities, the impact of institutional reforms like manager government, at large constituencies, and nonpartisan elections seem to have depressed not only citizen participation, but also governmental responsiveness. [17] Most officials elected under such systems are amateurs and will rarely rise to higher office. In a particularly original and forceful analysis, Kenneth Prewitt has described how the widespread norm of "volunteerism" among local elected officials reduces their fear of electoral reprisal and hence their accountability.[18] Moreover, if our arguments about the relative autonomy of the bureaucratic process in formulating decision-rules are correct, electoral reprisals against the "top dogs" are not necessarily efficacious in fundamentally altering the allocative process of urban governments. For all these reasons, and more, the electoral accountability model probably does not fare very well in explaining how the wants-needs-tastes-preferences of the citizenry get translated—if at all—into public policy.

Traditional democratic theory has never been fully accomodated to the realities of the bureaucratic city-state. Its focus has long been on the electoral side, while the bulk of Lasswellian politics in the city—who gets what how—takes place in other sectors. But bureaucratic theorists have tried valiantly to make democracy and bureaucracy compatible. One response of the democratic theorists has been to try to make bureaucracies *responsible,* that is, subordinate, to elected officialdom. Others would make the bureaucracy more *responsive,* or make them exhibit more concern, compassion, or committment to the clients they are supposed to serve. And some would make them more *representative,* i.e., more *like* the constituents they are supposed to represent.

Each of these approaches to more democratic bureaucracy has manifestations in contemporary urban politics. The criticism of bureaucracies by Lowi, who calls them the "new machines" and "relatively irresponsible centers of power," implies a desire to rectify the imbalance between elected or mayoral authority and bureaucratic authority. Much of the concern for "street level bureaucrats" is geared toward making them more responsive to the alleged needs of the poor, or the suspect, or the powerless. And the efforts to increase the ratio of minority police officers, social workers, or librarians entails a desire to make bureaucracies more representative. Each of these standards implies, of course, the inevitable tradeoffs with the others. Making bureaucracies more responsive to their clients would probably make them less responsive to higher authority. Making bureaucracies more representative does not make them more responsive, as the hostility to black police by black citizens indicates.

Also needed is a perspective on bureaucracy which evaluates them in terms of the traditional wants-needs-tastes-preference theory of democracy. Bureaucracies can be demand-processing institutions, just like elections. The unit of demand-making in the electoral system is the vote, while the analogous unit of demand-making in the bureaucratic system is consumption. One participates in elections by voting; one participates in the bureaucratic system by consuming public goods and services. It makes no sense to conceptualize the former as an expression of wants-needs-tastes-preferences, but not the latter. If participation in politics means registering demands about the size and scope of the public sector, rather than merely an exercise in the choice of candidates, people participate in politics through the act of service consumption and not merely by appearing on election day. If there is any doubt about which of these forms of participation is more important to the citizen—as opposed to the political scientist—we need only observe that the number of people voting is but a fraction of the number of people consuming public services.

Consumption as a mode of registering demands may actually offer advantages over voting as a means of registering demands. Elections are discreet, infrequent and somewhat artificial in being divorced from the mainstream of day-to-day existence. Moreover, elections represent a one-shot opportunity for inputting demands, while consumption permits a wide range of options. In statistical language, elections are a dichotomous variable while consumption is an interval variable. Generally, in statistical analysis, we prefer interval to dichotomous variables because of the probability that they will explain more variation. In elections, one votes or one refrains; one votes either for Jones or Smith. Variance in consumption is far greater. One may use a park daily or not at all; one may call the police frequently or never; one may ignore the public library, or virtually convert it into a private literary preserve. To put it another way, the inherent limitation on elections as a demand-registering device is the "one-man-one-vote" rule. Once given, a vote cannot (legally) be doubled or redoubled. Not so the consumption option, where citizens can cast multiple "votes" for a service by queuing up for it again and again. In the marketplace of public services, the citizen has multiple votes, at the ballot box only one.

Why, therefore, is bureaucracy so widely accused of being unresponsive to citizen demands? The market mechanism in the private economy is supposed to provide a mechanism (the famous "invisible hand") by which the production of goods and services is regulated. Part of the difficulty, we shall argue, stems from the very imperfect ways in which bureaucracies measure demands and consumption. Since bureaucracies are not selling a product, but merely delivering it, there is little incentive for them to know what they are doing. But an even more important reason why bureaucracies do not respond well to demands is that they are monopolies. And monopolies can afford to ignore demands better than competitive organizations.

HOW BUREAUCRACIES MEASURE DEMAND

We suggest that consumption is the potentially potent engine of demand-making in bureaucratic governance, just as elections are in political decision-making. That bureaucracies are roughly cognizant of changes in demand patterns can be shown. What they use may be described as a "body count" approach to consumption measurement. To urban bureaucracies, "performance" means counting consumption units. Every bureaucratic agency has its own consumption data, collected if for no other purpose than to include them in budget requests. The following "body counts" are taken from the various departmental reports and budgets of San Antonio public bureaucracies:

Welfare division: average yearly number of eligibility determinations for federal food stamps.

Rabies control division: dogs admitted to facility.

Environmental health division: square feet of water drained; birds bled for encephalitis control.

Streets division of public works: street sweeping collection, tons.

Parks department: total acres of parkland.

Delinquent tax office: lawsuits filed.

Fire department: alarms received and transmitted.

Library: book circulation.

Homemaker services: families served.

Police department: investigation of reported cases.

Streets: general patching, cubic yards.

These constitute typical output measurements of public agencies. What is striking about them in the first instance is that none of them contain a denominator. Being no more than raw consumption or delivery measures, they are roughly analogous to the famous Viet Cong body counts. The inevitable problem of those measures was whether body counts reflected changes in the number of VC engaged in guerilla activities (i.e., a change in the denominator) or changes in the kill ratio (i.e., a change in the ratio of the numerator to the denominator). But unless some denominator is found, raw demand or raw input measures are not very helpful in identifying changes in citizen wants-needs-tastes-preferences. Raw consumption measures exhibit the fatal flaw of assuming a direct relationship between outputs and wants-needs-tastes-preferences. This is an unseen hand indeed.

Moreover, raw rates of consumption or delivery tell us little—or tell bureaucracies little—about actual citizen preferences. There is a reciprocal relationship between some "pure" or "true" distribution of service preferences and the actual services make available. People's tastes for services are partially a function of the services they are aware of. The actual distribution may either enhance or inhibit consumption, and thus alter "taste." People who are suddenly given a satisfactory neighborhood library may change their reading habits. The opposite relationship holds as well. When services are seen as unsatisfactory, people's "taste" for them may diminish, inhibiting service consumption. The principal reason given, for example, for not reporting crime to the police, and thus registering demand through the consumption route, is the belief that the police could not help anyway. In such a case, need or preference is thwarted by perceived low quality of service. We have already shown a positive relationship between the accessibility and quality of public services and their consumption in our discussion of the distance-quality-consumption nexus.

Consequently, measuring public service consumption or delivery as an indicator of citizen preference is tricky business. People might have different preferences if the services to which they have access differed. Without a denominator—some measure of need, demand, or per capita equality—the use of raw service output data is not very useful except in justifying budget expansion. Add to that the reciprocal relationship between service availability and consumer taste, and the inferential difficulties of service output data are magnified. These problems relate to the ability of bureaucracies to *register* demands accurately. The monopoly problem relates to the bureaucracy's ability to *respond* to demands even when they are registered.

Service Bureaucracies and the New Monopoly Problem

THE NEW MONOPOLIES

Urban bureaucracies providing public services are, after all, monopolies. This truism is more or less descriptive, although not perfectly so. Lowi has observed that every public function has its private counterpart,[19] including private garbage pickup, private hospitals, private security forces, and even private postal systems. It is safe to say, though, that if urban public service bureaucracies are not strictly monopolies, they behave as if they were. Actually, the monopolization of urban services is a relatively recent development in the United States. Most now-public services were initially private. These included schools, fire protection, libraries, police protection, and public utilities. The demand for monopolization of public services was part of the great crusade of the same urban reform movement which spawned nonpartisanship, council-manager government, civil service reform, and bureaucratization. This concern of the reformers is not now often remembered largely because public monopolization of most services was, until recently, thought to be unassailable. But it was once a very hotly contested issue indeed. In 1899, Edward Bemis edited *Municipal Monopolies: A Collection of Papers by Eminent Economists and*

Specialists. There are, he said, "three great questions which now confront us: shall we have public regulation or public ownership and regulation? If the former, what shall be the nature of the regulation? If the latter, what are the dangers to be avoided?"[20]

One might think today that the municipal monopoly question is settled, but "it is an easy leap to the unwarranted implication that public goods paid for by the public through payments to the tax collector must be provided *to* the public *by* a public agency, *through* public employees."[21] Steven Savas has codified and popularized the attack on the monopolistic position of the bureaucracy, contending that

> the inefficiency of municipal services is not due to bad commissions, mayors, managers, workers, unions, or labor leaders; it is a natural consequence of a monopoly system. *The public has created the monopoly, the monopoly behaves in predictable fashion, and there are no culprits, only scapegoats.*[22]

EXIT, VOICE, AND MONOPOLY

The ultimate limitation of the bureaucracy in registering and responding to demands is that monopolies do not have any strong incentives to be responsive to their clients. There are, of course, different kinds of monopolies, but Albert O. Hirschman is especially fearful of one kind:

> But what if we have to worry, not only about the profit maximizing exertion and exactions of the monopolist, but about his proneness to inefficiency, decay, and flabbiness? This may be, in the end, the more frequent danger: the monopolist sets a high price for his product not to amass super-profits, but because he is unable to keep his costs down, or more typically, he allows the quality of his product to deteriorate without gaining any pecuniary advantage in the process.
>
> In view of the spectacular nature of such phenomena as exploitation and profiteering, the nearly opposite failings which monopoly and market power allow, namely, laziness, flabbiness and decay have come in for much less scrutiny.[23]

This characterization of monopoly is precisely Savas' fear, namely, that urban public service monopolies have allowed "the quality of [their] product to deteriorate."

In fact, some evidence has emerged supporting the "flabby monopoly" hypothesis. This evidence, fragmentary though it is, suggests that public monopolies provide lower quality services at higher costs than private competitors. The most extensive and useful analyses come from the areas of fire and sanitation. All-brandt, for example, studied the cost effectiveness of private fire protection contracting in Scottsdale, Arizona versus similar public fire departments, and found the latter wanting.[24] It costs the city of Chicago $1.23 to read one water meter, compared to 27.5¢ in Indianapolis which has a private-firm contract for meter-reading services.[25] Savas recounts the grim details of public versus private sanitation services in New York City. It costs the city $39.71 per ton to collect garbage, private firms only $17.28. Specifically, in Douglastown, an area of the city with single-family dwelling units, the city picks up garbage twice a week at curbside at an annual unit cost of $207. In nearby Bellerose, similar but outside the city, a private firm provides thrice-weekly pickups from backyards at an annual unit cost of $79. There is some irony, he notes, in the fact that capitalist Wall Street uses public garbage collection, while socialist Belgrade contracts out garbage collection to private entre-preneurs.[26] Savas has also studied the costs and efficiencies of private versus public garbage collection services in American cities. His data on 2060 American municipalities suggests that size of unit is an important factor in the relative cost-efficiency of public or private collection. But in cities over 50,000 contract collection is clearly more efficient.[27]

Citizens are unlikely to be aware of such evidence on the cost of municipal monopoly. What citizens possess is a vague feeling about the quality of service available to them. A rough judgment of "good" or "bad," and "better" or "worse" is about what one can expect. There is a smattering of evidence on citizen assessments of the quality of municipal services. Some of it is derived from the Urban Observatories' ten-city study of citizen attitudes. The pooled data in response to a "money's worth" question is found in Table 6.2. The evidence there suggests that the nay-sayers out-weigh the yea-sayers by a three-two margin. As one might expect, there are some differences by social class, but even among the highest status class, only a bare majority (54 to 46 percent) think that they are getting what they are paying for. Because these data come from a single point in time, they do not provide any evidence on the deterioration process. The closest approximation is a question concerning the quality of city government. These data are presented in Table 6.3.

Citizen-consumers were queried on their perceptions of changes in city government over the past half decade. Unfortunately, the question was not service-specific since it did not ask in what way government was worse (e.g., corruption in city hall or poor services) nor did it permit the registering of the option "worse." Even

Table 6.2: Responses by Socioeconomic Status to Question Concerning Whether You Get Your Money's Worth In City Services*

| Money's Worth | Socioeconomic Status | | | | | |
	Upper	Upper-Middle	Lower-Middle	Upper-Lower	Lower	Total
Yes	54.3% (134)	52.7% (289)	43.9% (282)	36.4% (550)	27.7% (148)	40.2% (1402)
No	45.7% (113)	48.3% (270)	56.1% (360)	63.6% (959)	72.3% (386)	59.8% (2088)
Total						100.0% (3491)

*The source of these data is the ten-city study of citizens attitudes toward urban taxes, services, and government performance. I am indebted to Professor Robert Wrinkle and to the Albuquerque Urban Observatory for making these data available.

Table 6.3: Responses by Socioeconomic Status to Question Concerning Whether City Government in the Last Five Years Has Gotten Better or Stayed the Same*

| Evaluation of City Government | Socioeconomic Status | | | | | |
	Upper	Upper-Middle	Lower-Middle	Upper-Lower	Lower	Total
Better	42.8% (92)	43.4% (193)	39.3% (193)	38.3% (426)	39.4% (146)	39.8% (1050)
Same	57.2% (123)	56.6% (252)	60.7% (298)	61.7% (687)	60.6% (225)	60.2% (1585)
Total						100.0% (2635)

*See note to Table 6.2.

with these limitations, we can assume some perception of stagnation if not deterioration since the ratio of persons picking "the same" to those who chose "better" is just about three-two regardless of social class. It might be added that during this period, city governments were increasing their expenditures faster than inflation, and therefore, the fact that services did not appear to get better could realistically be viewed as retrogression. This increase in budget may also account for citizens feeling that they are not getting their money's worth.

The monopoly problem is only a part of Hirschman's larger concern for the uses of "exit" and "voice" by customers faced with deteriorating firms. Though his analysis is focused on the private sector, its logic fits nicely the public sector as well. Exit options involve the customer's choice of a different firm, while voice is protest in some form to the "management." Generally, exit is economic behavior and voice is political behavior.[28] In popular parlance, exiters would "rather switch than fight"; voice exercisers "fight rather than switch." Hirschman's theory so nicely dovetails with the concerns of Savas and others for municipal monopoly that it deserves exposition in the context of urban service delivery systems. We suggest the relevant lines of option as depicted in Figure 6.1. Citizens, let us suppose, are dissatisfied with the quality and/or quantity of urban public services. In Hirschman's schema, there are only three options: exit, voice, and loyalty. Each is depicted there, together with subtypes of the exit and voice options.

THE "EXIT" OPTIONS AND MONOPOLY SERVICES

Now if urban public bureaucracies were literally monopolistic, and were producing a service for which demand were relatively inelastic, there would be no exit option whatsoever. If there were no other way of achieving the blessing of security, of garbage pickups, or recreation, or education, then exit would be a moot point. Urban public bureaucracies are not monopolies because they literally sell the only services in town, but *because they do not respond well to competition.* There are two modes of exit for those faced with the deteriorating monopoly.

The first one recalls the Tiebout hypothesis (see pp. 00-00 above) about "voting with one's feet." Because there is a multiplicity of governments within the metropolis, the rational economic man, seeking to optimize his tax-service mix, picks the "right" community in which to locate.[29] This model assumes that families and firms move about the metropolis in order to obtain a desired ratio of taxes and services. Although obviously limited by the realities of cost-space friction, the rational economic man theory rests upon familiar assumptions about economic man, behaving rationally, counting costs and marginal utilities. The assumption that "the citizen-consumer votes with his feet as he transfers his loyalty from some municipal Macy's to some municipal Gimbel's"[30] has underpinned some elaborate defenses of metropolitan fragmentation, on the grounds that optimality of choice is provided.[31] The mobility exit means simply that the customer of a deteriorating monopoly may cast about for a "better monopoly" elsewhere. In San Antonio, if he lives in any of the "city" school districts—Edgewood, San Antonio Independent, or Harlandale—he might well consider trying the Alamo Heights School District. If he had moved from, say, Edgewood to Alamo

Figure 6.1: Exit, Voice, and Loyalty Options of Consumers Faced with Perceived Deterioration in the Quality of Public Services

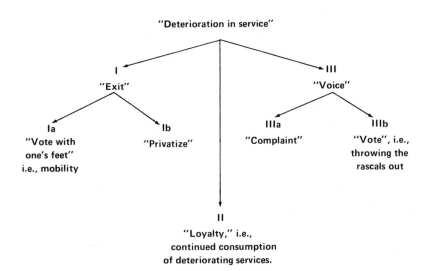

Heights around 1972, he could have more than doubled (from $248 to $558) the amount spent on his child at school. He could also have secured, for good measure, improvements in other public services. Such an exit is a literal one. But by continuing to purchase services from public monopolies, it essentially trades a deteriorating monopoly for a more productive one.

There is another form of exit, more important but less well-recognized, than "voting with one's feet." Exit need not be wholesale. Consumers do not always have to remove themselves to other communities to escape deteriorating monopolistic services. What they can do, increasingly, is buy out of the public sector into service provisions of the private sector. Instead of opposing such exits to contending firms, monopolies may even welcome the unloading of potentially troublesome customers. "Those who hold power in the lazy monopoly," Hirschman observes, "may actually have an interest in *creating* some limited opportunities for exit on the part of those whose voice might be uncomfortable."[32] The backsliding bureaucracy is relieved not only of some mouths to feed, but particularly of those most likely to cause it trouble. This is analogous to a bureaucratic decision which Savas has called "load shedding." Thus, instead of revivifying monopoly, competition may permit it to remain flabby and complacent. Indeed, one of the most important but underrecognized developments in American cities has been the increasing privatization of hitherto "public services." In the theoretically most "public" of all urban services, law enforcement, there are now twice as many private security personnel as public police officers. In Houston at least, the prevalence of private security patrols in upper-income neighborhoods is a rationale for police administrators to concentrate more resources in low-income areas.[33] In the schools as well, not only in the South but in the North, private school enrollments in some areas amount to 20 to 30 percent of totals. This could be attributed solely to racial motives were it not for the fact that analogous trends of privatization exist in other, less racially sensitive areas such as sanitation.

The notion of exposing government to competition has a rather heretical ring to American ears. It is naturally opposed by public service unions and employees. In fact, it was a concern over wage escalation and public employee strikes which led the British *Econ-*

omist to editorialize about "Let's Reprivatize." They comment that

> Britain's present round of wage inflation has been led by wage sur-
> renders in a public sector where productivity is unmeasured but often
> very low. In the United States, when garbage collection is handled by
> private contractors in parts of the west of the country, there has
> been ... evidence of costs being cut by 2/3 when *seedy municipal
> monopolies* ... are at last subject to competition. ... The proper way
> to keep down the costs of government spending will be for some future
> British government to brandish competition through threats of repri-
> vatization in the most surprising places.[34]

Small wonder that municipal employeers and teachers have re-
sisted contracting out, of which the school voucher scheme is the
most notable example.

The difficult empirical question, of course, is whether such
"threats of privatization" would really make bureaucracies more
efficient and responsive, or whether they would welcome the
opportunity to "load shed" their most troublesome customers.

THE "VOICE" OPTIONS

Voice is the political response to deteriorating services. Like
exit, it can take two forms. In the first, "complaints" are directed
to "management" as requests for service correctives. Unfortunately, the axiom that "you can't fight city hall," seems to be
deeply embedded in urban citizenry, and the exercise of the
complaint option is far more limited than one might expect.
Eisenger studied the pattern of citizen contacts, including but not
limited to complaints, with governmental officials in Milwaukee.
Only a minority of his sample (33 percent of whites and 11
percent of blacks) had made any recent contact with public
officials and only a minority of that minority had lodged complaints. Most of the contacts (over 80 percent) were directed at
elected officials rather than at bureaucrats. The ability of the
complaint process to function as a feedback or performance measurement system, he argued, was quite limited. Whatever their
target, "as a source of information, a tabulation of these complaints and pleas for help or service would provide little aid for

officials responsible for assessing the performance of particular municipal services or agencies."[35]

The complaint process need not be individualistic. When collectivized, we refer to the complaint process as the formation of "protest groups." What with all of the discussion of the deterioration of urban services in the popular press, it is remarkable that the size of such organized complaint groups is so small. The pooled data from the Urban Observatories' ten-city study of citizens and city halls shows that only 14.3 percent of respondents claim membership in "an organization working on city problems." Naturally, such membership varies by social class, with 26.0 percent of the upper class, but only 5.7 percent of the lower class claiming membership.

The conditions under which such protest groups will be successful are rather limited. Lipsky analyzed the rent strikes in New York City and concluded that rather considerable organization, quite specific goals, and access to third-party influentials (the press and the more established community leadership groups) was essential for protest to be effective.[36] There is also a good chance that the socioeconomic status of protest groups is related to the efficacy of their voice. We earlier recounted the distressing story of Newark, where a black neighborhood failed to secure a traffic light after months of petitioning and hundreds of signatures, while a white neighborhood got a similar request through in jig time.[37] There are parallels in the efficacy of protest groups in all cities, including San Antonio. For years, the city has run freeways through lower-status neighborhoods without provoking efficacious opposition, but the possibility of putting the North Expressway through a small number of expensive homes in the Olmos Park area engendered enough opposition to tie up the project for more than a decade.

The other voice option is "throwing the rascals out" through the electoral process. As we reach elections as modes of exercising voice, we come round again to where we began. We already suggested certain limitations of the electoral system as a demand-registering system. Any standard checklist could include low turnout, limited choice (especially in a place like San Antonio where a single elite is so dominant), the "volunteerism" of local elected officials and the limited opportunity for electoral reprisal, the impact of nonpartisanship, and the like. But there is a further problem with elections as a "voice" to change the delivery of urban services. Changing the big fish rarely changes the small ones. Mayors and council members come and go; bureaucracies and their decision-rules are more permanent. We have not argued that the council's role in allocative decisions is nil. Where a new library should be built, what route a highway should take, how revenue sharing funds should be allocated, and the like are all council-dominated questions. But the day-to-day allocations of valued urban services is more accurately called a bureaucratic phenomenon. The "reach" of elections as voice is shorter than assumptions of traditional democratic theory imply.

THE OTHER FACE OF SERVICE INEQUALITY: THE HIGH COST OF EXIT AND THE INEFFICACY OF VOICE

The problem of equality and urban public services has two faces. The first, that municipal governments may disadvantage the underclass through differential provision of public services, is the more common focus. The other face of service inequality, however, may be more important. It arises from the semi-monopolistic character of public service delivery systems, and the gist of it is this: For the urban underclass, the exit option may be costly but the voice option inefficacious. Put into propositional terms, the argument is as follows:

1. The higher the socioeconomic status of an individual or group, the greater the ability to exit, either by privitizing services or by "voting with one's feet";

2. The higher the socioeconomic status of an individual or group, the greater the efficacy of voice options, including complaining or conventional political participation.

There is even a corollary proposition concerning the additional effects of minority status:

3. Minority group status further reduces the ability to exit (in either of its forms) and the efficacy of voice (in either of its forms).

The message of these propositions is that exit and voice are costly political decisions for the individual. Neither comes cheap.

Escaping a deteriorating public school system, or a high crime rate, or unsightly neighborhoods involves either moving or securing comfort and convenience from the private sector. Some groups are simply better able to take advantage of these options than others. The ease of exit is roughly proportional to affluence. When racial barriers also exist, minority groups suffer a sort of double jeopardy in attempting to exit an unresponsive monopoly. It is remarkable, for example, that outside-central-city (suburban) areas grew by 28 percent between 1960 and 1970, but their black proportion remained exactly constant at 4.8 percent.[38] Despite a substantial narrowing of the black-white income gap in this period, exit remained more improbable or more costly for minority families. To some degree minority status displaced low affluence as a constraint on exit.

Two proponents of maximizing exit options, Bish and Ostrom, put the problem like this:

> Local governments in a democratic federal system are not pure monopolists. Citizens do have alternatives available to them which can generate competition. Competitive pressures can occur when people seek recourse by electing different officials, voting with their feet, using private alternatives or . . . other levels of government. However, all of these options have costs, and many of these costs may be hard for low income and minority citizens to pay. The cost of making their demands known to city hall may also be too high for them to pay. *They may have little option but to bear the burden of poverty until alternative structures can be devised to facilitate a redistribution of income and to provide governmental units that are more responsive to the demands of the poor.*[39]

This is an extremely hard-boiled defense of the exit option, exposing quite candidly its correlation with affluence. Basically, it holds that until economic redistribution or alterations in political power are achieved, the poor may find exit unavailable and voice inefficacious, leaving them no option but "loyalty" to deteriorating public sector services.

The other face of service inequality, then, pertains to the interfacing of the public and private sectors. This study of San Antonio—as well as studies of other American cities—has found less support for the underclass hypothesis than is generally assumed. The equal delivery of services to various neighborhoods, however, does not capture entirely the service problems of the urban underclass. Some of these problems are fundamentally structural, and relate to the range of options available in response to urban service monopolies and their outputs. In some cities, the problem may not be one of whether services are *equally delivered,* but one of whether services are *equally bad.* If services are equal, but deteriorating, the underclass retains the fewest options. That is the other face of service inequality.

Conclusions

This chapter has exhibited a dual thrust. First, it has attempted to buttress the "variance explained" in urban services by introducing the bureaucratic decision-rule explanation. Because the various sociospatial explanations of service distribution seemed to fit better a "variance discounted" perspective, we identified the minutiae of bureaucratic decisions as explanations for service outputs, though admittedly at a descriptive rather than a quantitative level. We treated consumption of public servies as analogous to elections, i.e., as a form of citizen demand-making or as participation in the political system on the output side instead of the conventional input side. The analysis of the deficiencies of the electoral system in registering and responding to citizen demands is now a commonplace in urban political science. We have also tried to detail the deficiencies of the consumption process as citizen demand-making. Service consumption, though potentially very significant, suffers several inherent obstacles. The most elementary is that bureaucracies do not try very hard to register and codify consumption-as-demand. The "body count" approach to measuring raw consumption provides little useful information. A small revolution in service delivery could be wrought by no more than the requirement that service agencies identify appropriate denominators, indicating need, per capita consumption or size of target groups. But, more significantly, the reciprocal relationship between service quality and wants-needs-tastes-preferences means that variation in service quality may be a function not only of changes in demand but of changes in the quality of services.

Unraveling these joint influences to sort out the "true" distribution of demands would be quite difficult.

Many of the deficiencies of bureaucracies as demand-processers relate to their monopolistic character. To the degree that public service bureaucracies represent Hirschman's "lazy monopoly," they suffer as demand-processers because invisible hand competition does not spur them to reform. Urban service monopolies are rarely, however, "pure" monopolies. From private septic tanks to private security patrols to private recreational opportunities, there is always "competition" and hence an exit option to those who can afford it. Thus, the other face of inequality in public services is that the exit option is variably open to those with different incomes or racial attributes. In the final analysis, this capacity of some groups to exit—either literally or by privatizing service purchases—may be more important than conventional wisdom about the underclass hypothesis.

NOTES

1. Henry Pirenne, *Medieval Cities* (Princeton, N.J.: Princeton Univ. Press, 1939), pp. 206-207.

2. Gideon Sjoberg, Richard A. Brymer, and Buford Ferris, "Bureaucracy and the Lower Class," *Sociology and Social Research*, 50 (April, 1966), pp. 325-337.

3. Frank S. Levy, Arnold J. Meltsner, and Aaron Wildavsky, *Urban Outcomes: Schools, Streets, and Libraries* (Berkeley: Univ. of California Press, 1974).

4. Wallace Sayre and Herbert Kaufman, *Governing New York City* (New York: Russell Sage, 1960), p. 405.

5. Theodore J. Lowi, "Machine Politics—Old and New," *The Public Interest* (Fall, 1967), pp. 86-87.

6. E. S. Savas and Sigmund G. Ginsberg, "The Civil Service: A Meritless System?" *The Public Interest*, (Summer, 1973), pp. 70-85.

7. J. David Greenstone and Paul E. Peterson, *Race and Authority in Urban Politics* (New York: Russell Sage, 1973), p. 215.

8. "Toward a Theory of Street-Level Bureaucracy," in Willis D. Hawley and Michael Lipsky, eds., *Theoretical Perspectives on Urban Politics* (Englewood Cliffs, N.J.: Prentice-Hall, 1976), Chapter 8.

9. Quoted in Mabel Walker, *Municipal Expenditures* (Baltimore: Johns Hopkins, 1930), p. 37.

10. See Aaron Wildavsky, *The Politics of the Budgetary Process* (Boston: Little, Brown, second edition, 1974); Ira Sharkansky, "Economic and Political Correlates of State Government Expenditures: General Tendencies and Deviant Cases," *Midwest Journal of Political Science,* 11 (May, 1967), pp. 173-192; and John P. Crecine, *Governmental Problem Solving: A Computer Simulation of Municipal Budgeting* (Chicago: Rand McNally, 1969).

11. Interview with San Antonio budget officer, June 14, 1972.

12. Eliot Ness, "Foreword" to Orlando W. Wilson, *Distribution of Police Patrol Forces,* publication no. 74 of the Public Administration Service (Chicago: Public Administration Service, 1941).

13. Kenneth Mladenka, "Serving the Public: The Distribution of Municipal Services," unpublished Ph.D. dissertation, Rice University, 1975.

14. Levy, et al., *op. cit.,* pp. 229-237.

15. Burt Pines, "Opinion Re Legality of Disparities," Office of the City Attorney, City of Los Angeles, June 3, 1975, pp. 2-3.

16. See the discussion and literature cited in Thomas R. Dye, *Politics, Economics and the Public* (Chicago: Rand McNally, 1966), pp. 16-18.

17. Robert L. Lineberry and Edmund P. Fowler, "Reformism and Public Policies in American Cities," *American Political Science Review,* 61 (September, 1967), pp. 701-716.

18. Kenneth Prewitt, "Political Ambitions, Volunteerism, and Political Accountability," *American Political Science Review,* 61 (September, 1967), pp. 701-716.

19. Theodore J. Lowi, *The End of Liberalism* (New York: W. W. Norton, 1969), pp. 44-45.

20. Edward Bemis, *Municipal Monopolies: A Collection of Papers by Eminent Economists and Specialists* (New York: Crowell, 1899), p. v.

21. E. S. Savas, "Municipal Monopolies versus Competition in Delivering Urban Services," in Willis D. Hawley and David Rogers, eds., *Improving the Quality of Urban Management* (Beverly Hills: Sage, 1974), p. 483.

22. Savas, "Municipal Monopoly," *Harper's Magazine,* (December, 1971), p. 55.

23. Albert O. Hirschman, *Exit, Voice, and Loyalty: Responses to Decline in Firms, Organizations, and States* (Cambridge, Mass.: Harvard Univ. Press, 1970), p. 57. For an interesting application of Hirschman's schema to urban politics, see John Orbell and Toru Uno, "A Theory of Neighborhood Problem-Solving," *American Political Science Review,* 66 (June, 1972), pp. 471-489.

24. Roger Allbrandt, "Efficiency in the Provision of Fire Services," *Public Choice,* 16 (Fall, 1973), pp. 1-15.

25. Chicago *Daily News,* (November 25, 1974), p. 13.

26. Savas, "Monopoly versus Competition," *op. cit.*

27. "Evaluating the Organization of Service Delivery, Solid Waste Collection and Disposal," Center for Government Studies, Graduate School of Business, Columbia University, October, 1975, mimeo.

28. Hirschman, *op. cit.,* p. 31.

29. Charles M. Tiebout, "A Pure Theory of Local Expenditures," *Journal of Political Economy,* 64 (October, 1956), pp. 416-424.

30. Norton Long, "Social Science and the City," in Leo Schnore and Henry Fagin, eds., *Urban Research and Policy Planning* (Beverly Hills: Sage, 1967), p. 245.

31. See, e.g., Robert L. Bish and Vincent Ostrom, *Understanding Urban Government* (Washington, D.C.: American Enterprise Institute for Public Policy Research, 1973).

32. Hirschman, *op. cit.,* p. 60.

33. Mladenka, *op. cit.*

34. April 12, 1975, p. 14.

35. Peter K. Eisenger, "The Pattern of Citizen Contacts with Urban Officials," in Harlan Hahn, ed., *People and Politics in Urban Society* (Beverly Hills: Sage, 1972), p. 53.

36. Michael Lipsky, *Protest in City Politics* (Chicago: Rand McNally, 1970).

37. Michael Parenti, "Power and Pluralism: A View from the Bottom," *Journal of Politics,* 32 (1970), p. 513.

38. William W. Pendleton, "Blacks in Suburbs," in Louis H. Masotti and Jeffrey K. Hadden, eds., *The Urbanization of the Suburbs* (Beverly Hills: Sage, 1973), pp. 172-73.

39. Bish and Ostrom, *op. cit.,* p. 31.

Chapter 7

THE LIMITS TO EQUALITY

*The law, in its majestic equality, forbids the rich as well as the poor to
sleep under bridges, to beg in the treets, and to steal bread.*

A. France

*Just as there is no point at which the sea of misery is finally drained, so,
too, there is no point at which the equality revolution can come to an
end, if only because as it proceeds, we become ever more sensitive to
smaller and smaller degrees of inequality.*

N. Glazer

There are policy problems connected with the delivery of urban
services, although their character is poorly suggested by the widespread "underclass hypothesis." We will argue in this concluding
chapter that our evidence from San Antonio is strikingly consistent with other analyses of urban service distribution in major
cities. While not denying some contrary evidence, we suggest that
the weight of evidence—albeit from an unsystematic sampling of
research sites—is definitive in showing that municipal governments
more often than not achieve a rough equivalence of service packages among their neighborhoods. When that rough equivalence is
violated, it is sometimes violated in a direction opposite to the
conventional wisdom about the underclass and city services.

This conclusion has significant implications for political, legal,
and administrative strategies regarding the urban services problem.
In particular, it suggests that it will ordinarily be exceedingly
difficult, using the normal methods and canons of evidence from
the social sciences, to demonstrate constitutionally-proscribed discrimination against the urban poor. To those who favor a strategy
of legal redress, this conclusion will be disquieting—almost as

disquieting as the recent reluctance of the Nixon Court to sustain any serious challenge to state policy on racial grounds. Legal strategies, however, involve a cost-benefit calculus. Our reading of both the evidence and the state of constitutional interpretation suggests that the costs are high and the probable benefits are uncertain.

There is no implication in our argument that the state of municipal services is satisfactory or that the poor are not perpetuated in their poverty by the policy choices of municipal governments. Indeed, those who are concerned to upgrade the status of the urban underclass would do well to favor more *inequality* in urban service distributions than probably now exists. The problem we specify stems not from the public sector's service delivery system, but rather from the nexus of a roughly egalitarian public sector and an extremely unequal distribution of private sector advantages. This we call the "other face" of service inequality.

Much social science analysis of policy issues ends not with a bang, but with a whimper. Evidence is developed and conclusions are drawn with little attention to their relevance for political, legal, or administrative options. Sometimes one solution is glibly advocated after a careful analysis has found another one wanting, shifting problems from one to another of Pandora's boxes.[1] Here we cannot exit with much of a bang, but we can at least avoid going out with a whimper. Social scientists now know enough about the distribution and delivery of urban services to suggest at least the context of choice about urban service allocation. We shall argue that equality is one parameter in that choice, but not the only one.

The Evidence: San Antonio and Elsewhere

ENCAPSULATING THE SAN ANTONIO EVIDENCE

Typically, social science proceeds by the progressive testing of commonplace hypotheses. More often than not, these hypotheses prove essentially true—whites do vote more than blacks, southern congressmen are more conservative than northern congressmen— and social scientists risk the charge of merely "verifying the obvious." Other elements of the conventional wisdom, however, are found wanting. Pupil-teacher ratios are unrelated to pupil performance, routine police patrols do not have a noticeable effect on crime rates, urban services are not strongly correlated to neighborhood income, and related arguments invite controversy over policy analysis and provoke charges of error. Social scientists, therefore, walk a very thin line between being "wrong" and being "trivial" in the estimation of their critics. This research began with one of those commonplaces, typified by Claude Brown's unscientific but forceful hypothesis that "Harlem was getting fucked over by everybody." Our evidence, unfortunately, does not extend to Harlem. The actual measurement of delivery patterns is a complex enterprise, and we have had to settle for evidence from a single city. We shall review in the next section the evidence from other communities. But our own evidence about San Antonio can be summarized in a few simple propositions:

1. The distribution of urban public services in San Antonio can be characterized as one of "unpatterned inequality."

2. Neither neighborhood ethnicity, nor political power, nor socioeconomic status are very satisfactory predictors of service allocations, casting doubt upon the underclass hypothesis.

3. To the degree that any attributes of neighborhoods are related to service delivery, their ecology (specifically, their population density and age of housing) is more closely related than any other attribute.

4. Older, denser neighborhoods are more proximate to public service facilities, and the quality of services there is roughly equivalent to other neighborhoods; hence, more service "discrimination" is suffered by residents of peripheral areas, beyond the present outreach of municipal service networks.

5. The pattern of tax assessment is also better explained by ecological than by underclass hypotheses, with homes whose assessments most closely approximate their "true value" (by our admittedly crude estimates) located in newer, less densely settled (hence peripheral) areas.

Introducing the bureaucracy into the service allocation milieu, we suggested that bureaucratic decision-rules are probably more critically linked to delivery than specific attributes of neighborhoods. Decision-rules are largely generated by agencies themselves, being a product both of precedential handling of like situations at all levels of the bureaucracy—which becomes, upon reproducing itself sufficiently, the process commonly called incrementalism—and of definitions of the situation invoked by members of the bottom echelon of the bureaucracy. Consequently,

> 6. urban public bureaucracies, through their discretion both to make delivery rules and to fit a particularistic claim to one of several rules, probably have more to do with the allocation of services than does the distribution of political power.

Even in a city remarkably conscious of its own "power structure"—the San Antonio *Light* runs an annual listing of its choice for the ten most powerful people and organizations in the city and compares its listing to a readers' poll on the same subject— variations in concentration of neighborhood power had little to do with variations in service outputs. Service decisions are administrative decisions.

EVIDENCE FROM ELSEWHERE

There is always a risk in overinterpreting evidence from a single city, whether it is New Haven, Atlanta, Oakland, Chicago, or San Antonio. We defended our choice of San Antonio, though, not only on practical grounds, but precisely because its uniqueness— Will Rogers called it one of America's four unique cities—offered an especially appropriate test case. Its concentrated power structure, the minimal suburbanization, and the heavy incidence of the urban underclass made it ideal for an examination of the underclass hypothesis. Still, it is only a single case. Our conclusions will be made more compelling if we can demonstrate that, with regard to its distribution of urban services, San Antonio is not at all unique.

Table 7.1 itemizes several major assessments of the neighborhood allocation of three urban services (parks, police, and libraries) in several communities. There are, of course, assessments of other service domains as well, and their findings are equivalent to those reported in our review-of-the-literature in tabular form. Represented there are mostly studies of large cities like New York,

Table 7.1: Selected Studies of Urban Service Distribution

Study, Place, Date	Dependent Variable	Findings
I. Libraries		
1. Blank, New York City (1969)	Distribution of libraries' expenditures, bookstock	No clear relationship to income or race of neighborhood
2. Mladenka and Hill, Houston (1977)	Distribution of library resources; distance	No clear relationship to income or race of neighborhood
3. Levy, Oakland (1974)	Distribution of library books	Clear, direct relationship between income of neighborhood and library quality
4. "Access to Public Libraries," various cities (1963)	Distribution of library facilities, books, etc.	Clear, direct relationship in most cities, between resources and race of area served, with white neighborhoods getting superior libraries
5. Martin, Chicago (1969)	Distribution of library resources	Clear direct relationship between income and race and quality, proximity of library
II. Police Protection		
1. Bloch, Washington, D.C. (1974)	Distribution of police manpower	No clear differences among different neighborhoods
2. Weicher, Chicago (1971)	Distribution of police manpower	Heaviest concentrations of services in wealthiest and poorest neighborhoods, least in middle-income neighborhoods
3. Mladenka, Houston (1975)	Response time by neighborhood	No clear difference by race or income of neighborhood
III. Parks		
1. Community Council of Greater New York (1963)	Park facilities	Favored higher-income areas
2. Gold, Detroit (1974)	Park facilities, expenditures	Favored low-income areas
3. Fisk, Washington, D.C. (1973)	Park facilities	Favored low-income areas
4. Lyon, Philadelphia (1970)	Capital expenditures on parks	Equal by neighborhood
5. Mladenka and Hill, Houston (1975)	Facilities, proximity	Equally distributed by neighborhood, with a slight tendency to favor disadvantaged neighborhoods

Sources:

American Library Association, *Access to Public Libraries* (Chicago: American Library Association, 1963).

Blank, Blanche D., et al., "A Comparative Study of an Urban Bureaucracy," *Urban Affairs Quarterly*, 4 (1969), pp. 343-354.

Bloch, Peter, *Equality of Distribution of Police Services: A Case Study of Washington, D.C.* (Washington, D.C.: The Urban Institute, 1974).

Community Council of New York (1963), *Comparative Recreation Needs and Services in New York Neighborhoods* (New York: Community Council of Greater New York).

Fisk, Donald, et al., *How Effective Are Your Community Recreation Services?* (Washington, D.C.: Department of the Interior, 1974).

Gold, Steven D., "The Distribution of Urban Government Services in Theory and Practice: The Case of Recreation in Detroit," *Public Finance Quarterly*, 2 (January, 1974), pp. 107-130.

Levy, Frank, et al., *Urban Outcomes* (Berkeley: Univ. of California Press, 1974).

Lyon, David, "Capital Spending and the Neighborhoods of Philadelphia," *Business Review* (of the Federal Bank of Philadelphia), (May, 1970), pp. 16-27.

Martin, Lowell A., *Library Response to Urban Change: A Study of the Chicago Public Library* (Chicago: American Library Association, 1969).

Mladenka, Kenneth, "Servicing the City: The Distribution of Municipal Services," Unpublished Ph.D. dissertation, Department of Political Science, Rice University, 1975.

_____ and Kim Q. Hill, "The Distribution of Urban Police Services," *Urban Affairs Quarterly*, 12 (1977).

Weicher, John C., "The Allocation of Police Protection by Income Class," *Urban Studies* 8 (October, 1971) pp. 207-220.

Chicago, Washington, Oakland, and Houston. Most of the court cases dealing with service discrimination have tended to come from small towns. While one could easily find exceptions in all areas, the cumulative weight of the evidence seems clear. Distribution studies are more likely than not to find either roughly equal or even compensatory patterns than to find discriminatory patterns. The service area where the evidence is most unclear is libraries. In Oakland, Chicago, and the cities investigated by the American Library Association's "Access Study," library distributions—largely because of the practice of allocating new books to high-circulation branches—favored advantaged neighborhoods.

The cumulated evidence on the distribution of urban services in larger cities suggests plainly that "objective," social science methodology is not likely to uncover vast differences in the quality of services to neighborhoods in larger cities. *Pockets* of discrimination can be found, but probably not *patterns* of discrimination. There are no doubt dozens of ways municipal governments discriminate against the urban underclass, including zoning decisions, the *treatment* of citizens by bureaucracies, and educational disadvantages. But the weight of evidence compels the conclusion that overt, measurable discrimination in the distribution in conventional city services has been overstated by anecdotal commentary and conventional wisdom. The plight of the poor in the city cuts deeper than service discrimination alone.

THE EVIDENCE OF DISTRIBUTION AND THE OTHER FACE OF INEQUALITY

If our review of the evidence is accurate, then the distribution of urban public services is not a function of the underclass hypothesis. Poor people there are in abundance in any city; but they do not universally receive the poorest public services. Rather, toward the end of the previous chapter, we suggested that there are two "faces" of the problem of inequality and urban services. The first is suggested by the underclass hypothesis, which holds that the poor, the powerless or racial minorities, are victims of service discrimination. It is this hypothesis we have tested in one city and found wanting. The other face, however, arises from the conjuction of a relatively egalitarian public sector and a distinctly inegalitarian private sector. Inequality of resources leads to variations in access to exit and voice options. The inaccessibility of exit and voice, in turn, reduces a citizen's ability to respond effectively to inadequacies in the quality of bureaucratic service production. Even when public sector inequalities are trivial, egalitarian issues of urban services are not solved. They will be solved only when there are vehicles for insuring equal probabilities of responsiveness and equal access to private sector alternatives to inadequate service provision.

Having said this, we have turned our original problem into a much more complex one. If the activities of urban governments systematically discriminated against lower-class citizens (as they so patently did in Shaw, Mississippi and no doubt elsewhere), legal remedies are in principle available. If, however, the evidence is otherwise, the roots of urban inequality must be sought in the nexus of public-private sector resource allocations. We suggested in Chapter 2 that service distributions could be described as *direct, compensatory,* or *equal* vis-á-vis neighborhood status. Direct allocations follow a "them that has gets" pattern; compensatory allocations advantage the least well-off in private sector resources; and equal patterns give proportionate shares to neighborhoods regardless of their income or status. When an equal pattern prevails, the law may be satisfied. But the disadvantaged may not be. The law may be content with equality in service delivery. Those who favor redistribution as a social value will support compensation.

We have, in other words, an answer to the problem with which we began. In San Antonio and elsewhere, the pattern of service delivery is roughly equivalent—with some notable exceptions in some cities—and reduces the first face of inequality to a less serious problem than some commentators believed. The larger problem of public services and inequality, however, is not resolved. If public services become more equal over time, especially if they become equally bad, it becomes ever more important to hold enough private sector resources either to exit or to privatize.

Services, Equality, and Other Values

If all we expect of urban services is that they be equally delivered, we will be relatively satisfied with the public service provision of most large cities (with some important exceptions). But we obviously expect more of public services than this. We

desire services not only equal from neighborhood to neighborhood, but equally *good*. Few will take comfort in the fact that
police nowhere respond promptly, that parks are in a state of
universal disrepair, and that all the city's streets are pockmarked.
Urban public services represent both intrinsic and instrumental
values. They are, first of all, a *service*. The political problems in
insuring *quality* of service outputs are very similar to those of
insuring *equality* of services.

Once some adequate level of quality has been secured, we can
proceed to evaluate services in terms of their instrumental values.
There are several evaluative norms for assessing public services.
Our analysis has been based upon an egalitarian norm, for few
standards are so universal in the law, in philosophy, and in the
social sciences. The danger, always present with universally revered
concepts like equality, is that it will be pumped clean of its
normative crispness. It will become, as one philosopher put it, a
word "so wide and vague as to be almost unmeaning." So long as
there are blatant and brutal discriminations in the allocation of the
public largesse, epitomized by the "Hawkins" case or by the
example of the road grader in one southern county which picked
up its blade in black neighborhoods, there will be a place for the
equality test of urban services. Yet an equality test alone is both
too strong and too weak. It is too strong because an excessive
fixation upon egalitarian premises leads to Glazer's paradox of the
"sea of misery which will never be drained," as we become
sensitive to the most miniscule deviations from the equality principle. It is also too weak because it squeezes out attention to other
values which public policies are expected to maximize. The relent-
less pursuit of equality will require tradeoffs with other, sometimes equally compelling, decisional principles.

WHICH TRADEOFFS?

The most elemental rule of modern economic analysis is that
one cannot maximize two values simultaneously. With fixed resources, we cannot have the most of both guns and butter,
Lyndon Johnson to the contrary. Decision-makers can never behave like Stephen Leacock's famous rider who flung himself on his
horse and rode off in all directions at once. As with any matter of
social policy, there are tradeoffs among contending distributional

standards of urban services. There are four principal goals which merit attention: equality itself, efficiency, citizen preference, and need. Only the first of these is explicitly a judicial test, but this alone should not give equality any special privilege as a normative standard.

A sound case can be made for the distribution of public services according to any of these tests. Need, for example, is rightly considered a slippery standard for policy which, like beauty, tends to exist primarily in the eyes of the beholder. The Supreme Court once refused, probably rightly, to review a holding that failure to meet educational "needs" of pupils did not present a judiciable issue.[2] In the social sciences, only Abraham Maslow's foray into the construction of a "hierarchy of human needs" has ever attempted seriously to incorporate the idea of need-fulfillment. Yet there is a weaker sense in which urban policy is clearly and defensibly infused with need tests. There is a rough correspondence of professional and lay opinion that some areas "need" more fire protection (airports, downtowns, areas of old frame houses) than others. The same is true with law enforcement services. Orlando Wilson, who contributed perhaps more to the professionalization of law enforcement agencies than anyone, argued that "police personnel should be distributed in proportion to the *need for police service.* The essence of the problem of distribution lies in measuring proportionate need."[3] The virtue of the need principle over any other is that it explicitly rationalizes deviation from egalitarian standards in the direction of compensating for private sector inequalities. In recreation services, for example, "people in low income families are likely to be less able to provide their own leisure-time activities than people in higher-income families. Because they are less mobile and have less amenities among their own resources, low income families on the whole have a greater need for community supported recreation services."[4] The deficiencies of need as a judicial stand do not temper its utility as a test of social policy.

Efficiency and citizen preference do not figure in a constitutional test of service outputs. If bureaucracies cannot measure efficiency very effectively, nor social scientists estimate orderings of citizen preferences, it would be unreasonable to expect courts to fill such vacuums. Here we set aside the more elaborate meaning of efficiency as Paretean optimality, simply because there are too many missing links between the theory of welfare economics and operating decisions about urban services. An important distinction can be drawn, however, between efficiency in welfare economics— efficiency writ large—and administrative efficiency—efficiency writ small. Administrative efficiency is securing the greatest output for the least amount of input. The recent explosion of concern for public sector productivity measurement incorporates this emphasis on administrative efficiency. New York City has experimented with improvements in productivity, although the power of municipal employee unions has blunted somewhat the full implementation of the system. The point, however, is that an efficient allocation of bureaucratic resources may also be an unequal one.

Shoup, Thurow, and others have argued forcefully that equity-equality tests and efficienty tests of law enforcement produce very different distributional patterns.[5] The argument can be generalized to other service domains as well. It would therefore be an inefficient distribution of police resources to equalize protection in every neighborhood regardless of whether the crime rate were high or low. Many crimes in high-crime neighborhoods would go unchecked, while a service "overkill" would appear in low-crime areas. Partly because of the inherent difficulties in determining the efficient use of public resources, courts have always given wide discretionary latitude to administrative determination of service allocation patterns. In "Gowan v. Smith," for example, it was noted that "the [police] commissioner is bound to use the discretion with which he is clothed... To enable him to perform the duties imposed upon him by law, he is supplied with certain limited means. It is entirely obvious that he must exercise discretion as to how those means shall be applied for the good of the community."[6]

Finally, citizen preference can determine urban service packages. In a democracy, the demands of citizens are presumably a touchstone of public sector allocations. Very little is known, unfortunately, about variations in the preference orderings for public services among urban neighborhoods. The Urban Observatories ten-city attitude survey about urban services did inquire about whether citizens perceived their neighborhood getting a fair share of services and whether they desired more or less service expenditures. Hoffman's analysis of these data uncovered the paradoxical finding that both groups which saw themselves as advantaged and those (especially minorities) which saw themselves as disadvantaged preferred bigger service packages.[7] Citizens who saw their neighborhoods as getting its fair share and no more were least likely to want more service outputs.

Two schools of though, which otherwise have little in common, have posited wide variations in neighborhood service preference and proposed restructuring urban governments to accomade these preferences. The "community control" literature advocated neighborhood governments responsible for basic municipal services to serve particularistic needs of (especially minority) neighborhoods.[8] During the 1960s, several halting efforts were made to actualize community control in New York, Detroit, and elsewhere. But municipal bureaucracies were successful almost everywhere in beating back such particularistic challenges to their universalistic and monopolistic hegemony. From a very different quarter, economists in the "public choice" tradition have defended the multiplicity of metropolitan governments because citizen-consumers can presumably select a municipal government which satisfies their optimal tax-service mix.[9] Whatever the rationale, the idea that citizen preferences should be directly tied to service decisions is not—or should not be—alien to democratic local government.

Neither need, efficiency, nor citizen preference, however, is likely to produce equality of service delivery. Need can be justified on egalitarian grounds, primarily because it provides unequal services to compensate for other sources of inequality. Efficiency may be associated with any distributional pattern. Citizen preference standards will produce equal service outputs to neighborhoods only with the unlikely assumption that all neighborhoods prefer identical service packages.

Four Standards and Four Strategies

Whether one prefers urban services to be judged primarily in terms of egalitarian, efficiency, need, or consumer preference norms, will determine in large measure political strategies. Alternative modes of political strategy are not incompatible, to be sure. But securing a firm constitutional interpretation holding that inter-neighborhood equivalency of services is mandatory will produce inefficient service patterns and delivery that poorly articulates needs. But throwing service decisions to the political process and expecting it to maximize consumer preference will give additional weight to groups commanding resources for effective exit and voice. What we want from urban services helps determine both institutional and political strategies.

THE COURTS AND THE EQUALITY STANDARD

Following de Tocqueville's dictum that "scarcely any political question arises in the United States which is not resolved, sooner or later, into a judicial question," urban public services have more frequently ended up in the courts. This has been especially true for education, but is increasingly the case with conventional public services. The "Hawkins" case and the efforts of civil rights attorneys to secure judgements against both southern and northern discrimination are exemplary. For those who favor the strategy, the equal protection clause looms as a powerful potential weapon to secure equal services to the urban underclass. If our arguments about San Antonio and elsewhere are correct, there will be relatively few open-and-shut "Hawkins" cases to be found. But where they exist, legal proceedings are a logical remedy.

Unfortunately, the Burger Court has begun to undercut some of the extensions of "substantive equal protection" offered by the Warren Court. The case of "Washington v. Davis"[10] offered the Court a chance to hold against a police examination which black applicants failed more frequently than white applicants. Emphasizing that "disproportionate impact . . . standing alone" does not violate the equal protection test, the Court concluded that "the invidious quality of a law claimed to be racially discriminatory must ultimately be traced to a racially discriminatory purpose." Even worse for the legal strategists of service equality, the majority singled out (in footnote 12) a list of cases—some of them on the docket yet to be heard—which it disapproved. Among them was "Hawkins." While it is frankly difficult to believe that there is no constitutional recourse for the citizens of Shaw, Mississippi, the decision bodes ill for those who would haul offending municipalities into court.

Even if the Court were likely to be more receptive to service challenges, there are significant difficulties involved in both measuring and fashioning remedies for service discrimation. Elsewhere, we have identified these as the problems of (1) the permissable range of variation; (2) the units of analysis problem; and (3) the input versus output equality problem.[11] The first poses the question of how much inequality should be tolerated as acceptable. When variations are as extreme as those in Shaw, Mississippi, one does not worry about whether they are intolerable. But minor deviations will exist even in a relatively equal pattern. A federal court in the District of Columbia mandated a range of variation no greater than ±5 percent in school-to-school expenditures. Yet the price of perfect equality is a high one. If the test of equality is applied too literally, it will become like Glazer's sea of misery which is never drained. Secondly, even assuming agreement on a permissable range of variation, to what sociospatial units shall the equality test be applied? The fourteenth amendment guarantees equal protection to individuals as such, and not to larger spatial aggregates. Yet arguments about neighborhood service denials point to larger units than households. If a larger unit is appropriate, then what particular aggregate is desired? The stipulation of a very small unit of analysis—say a block or even a census tract—would make equality of services a very expensive proposition because of diseconomies of scale. But the progressive enlargement of the unit raises questions about where to stop. Finally, guaranteeing "input equality" does not guarantee "output equality." Comparable investments in police protection or in street cleaning may still leave one neighborhood with much higher crime and dirtier streets than another. In sum, the "Hawkins" remedy—blatant though the facts of the actual case were—promises sounder remedies for extreme discrimination than for resolving equity problems as a whole.

BUREAUCRACY AND THE EFFICIENCY TEST

Despite the outpouring of sentiments against the "uncivil service," one had best made one's peace with the inevitable. One can, of course, believe that bureaucratic decision-rules are utterly venal, self-serving and wholly beyond redemption. If, however, bureaucracies in San Antonio somehow deliberately set about to design service allocations to favor particular groups of the powerful or the wealthy, they do a poor job of it. Our view is somewhat more benign, holding that bureaucracies have very little information about their own service outputs and very little incentive to develop any rationales for their decision-rules. In cities even as large as San Antonio, productivity measurement is so crude that it would take a back seat to that of a hamburger franchise. Unless urban governments improve their own measurement of what they are doing—and we have in mind more sophisticated measurement than the number of cubic yards of asphalt used in street repairs— there is scant hope of improvements in their distributive attention rules.

Those who endorse a "bureaucracy heal thyself" approach are bound to be disappointed. If one believes, however, that bureaucratic decision-rules ignore distributive consequences more because of ignorance than perniciousness, there is more prospect of change. Bureaucracies rarely know the distributive impacts of their decisions. The urban fiscal crisis has spurred efforts to improve productivity and delivery. One side effect is that productivity measurement requires data. Both officials and bureaucracies would be well-served to develop systems of neighborhood-based social accounts which would require measurement of batteries of services and needs in various neighborhoods. Were the conventional wisdom that bureaucracies always discriminate against certain neighborhoods verified consistently, it is true that bureaucracies would have much to hide. But Walsh puts the case for such an accounting of services at the neighborhood level as follows:

> Assuming willingness to seek at least some quantitative measures of minimally fair distribution . . . an information system is required to permit determination of existing current patterns of distribution. Districts could be identified for which each department of government could develop expenditure/per capita data and other measures of output. This is absolutely essential to permit areas of the city to determine what share they *are* getting. There is some evidence from a recent analysis of New York City . . . that the results might diverge considerably from popular impressions.[12]

Knowing where services go will at least provide citizen groups and policy-makers the opportunity to press for redistribution where the evidence warrants.

INSTITUTIONAL ARRANGEMENTS, POLITICAL ACTION, AND THE CONSUMER PREFERENCE STANDARDS

If consumer preferences vary, and are to be sovereign in determining service decisions, then neither the courts nor monopolistic bureaucracies can be sovereign. Both the courts and the bureaucracies apply universalistic standards of output. Those who support greater citizen control assume a considerable variability in preference orderings from group to group. Because groups are often spatially clustered within the city, neighborhoods exhibit corresponding variability in their preferences for public services. This is an empirical question. And, unfortunately, little is known about inter-neighborhood variation in preference for police, fire, recreation, school, street, or other services.

There are several ways in which consumer sovereignty can be maximized. A politics-makes-strange-bedfellows coalition of liberal advocates of community control and conservative public choice theorists favor the option of a *large number of small governments*. Those citizens inside a government can more easily utilize voice to insure their service preferences. Public choice theorists, like Tiebout, Bish, and Ostrom, add that those who object to the service preferences of their government can always "vote with their feet."[13] Either way, service distribution will probably be unequal, i.e., not identical, from one jurisdiction to another. They should, however, correspond more closely to citizen preference. Public choice theorists take the private marketplace model as applicable to the public service sector. It is the alternatives and competition offered by numerous governments which ultimately erodes monopolistic bureaucracy.

Along the same lines, it is also possible to create competition to public service monopolies by privitizing hitherto "public" services. If one cannot make bureaucracies more responsive to consumer preferences by hortatory means, then let them compete with one another. Voucher systems in the schools are the most obvious outcropping of this mode of institutional restructuring. Savas' evidence that private service contractors frequently provide better, yet cheaper, services gives sustenance to the argument.[14] One difficulty, however, is that private sector inequalities now seem far more pervasive than public sector inequalities. Demonopolizing and privatizing public services might exacerbate inequality at the expense of efficiency and citizen preference. Advantaged groups have always had the option of private service providers. It has been the poor who have most relied upon the public sector.

THE PROBLEM OF NEED AND THE
OTHER FACE OF INEQUALITY

The fact that the poor are more dependent upon the public sector casts doubt upon strategies which would set them at the mercy of exit and voice strategies, where they are at their weakest. This dependence of the poor on public resources is widely known, yet often underestimated. Browning estimates that the average poor family in 1973 received sufficient in-kind transfers to place its actual income at 130 percent of the poverty line.[15] Whether the creation of public sector dependency is the wisest response to the needs of the poor goes beyond our present concerns.

At the heart of our concern throughout, however, has been the proposition that urban public services constitute "real income," as distinct from pecuniary income. Indeed if service outputs are distributed with relatively little class-related variation, as they are in San Antonio, there is very little redistribution of real income being accomplished at the local level. Whether more redistribution is desirable is itself a political issue (or nonissue) of considerable magnitude and complexity. Even if it were settled in the affirmative, it does not follow that public services are a logical vehicle for achieving such redistributive objectives. It is possible that income reform alone would permit greater exit and voice by the poor. Why poor neighborhoods should be saturated with public services instead of simply being assured more jobs or more money is not self-evident. Services, as we said, are instrumental values, but they are also intrinsic ones. Saturation of poor neighborhoods with public services would increase their public sector dependency and settle a priori the issue of what the poor "need." Perhaps the poor would settle for the same opportunities for exit and voice available to advantaged segments of the urban community. Exit and voice are, in the American city, strongly correlated with the other face of inequality, the distribution of private sector resources.

There is another meaning of "need" in service-allocation standards, one more common sensical. Some areas of the city simply exhibit greater need for police protection (where traffic or crime rates are high) or park space (where multifamily dwellings predominante) or fire protection (where old frame units are clustered). These considerations mitigate against the development of any universalistic standards of service distribution.

Conclusion

Throughout, we have focused on "who gets what" within the urban community. Our principal research hypothesis—that the urban underclass was seriously shortchanged in the public sector—proved to be either overstated or simply wrong. Still, the problem of urban services and equity are not resolved by discovering that conventional wisdom erred. The delivery of urban services involves other values than equality narrowly conceived. Strong cases can be made for utilizing equality, consumer preference, efficiency, or need as contending standards for service delivery. Urban governments cannot maximize all those values at once, and tradeoffs are inevitably made. Our thesis here has been that tradeoffs are better made consciously than unconsciously, as decisions instead of nondecisions. What we expect of urban services determines our response to the egalitarian mandate. There are limits to equality as a standard, particularly as a legal standard. Some of those limits inhere in the paradoxical nature of the egalitarian precepts. Others stem from the need to trade off equality against other values. But whatever our preferences, the question of urban public service is too important to be left to the service producers alone.

NOTES

1. In our view, one such glib solution is Christopher Jencks' suggestion that income redistribution is the only appropriate response to the deficiencies of schooling. See his *Inequality: A Reassessment of the Effect of Family and Schooling in America* (New York: Basic Books, 1972), pp. 263-65.

2. *McInnis v. Ogilvie*, 394 U.S. 322 (1969).

3. "The Distribution of Police Patrol Force," publication no. 74 of the Public Administration Service, Chicago (1941), p. 5, italics added.

4. Edwin Staley, "Determining Recreation Priorities: An Instrument," *Journal of Leisure Research*, (Winter, 1969), p. 70.

5. See, e.g., Carl S. Shoup, "Standards for Distributing a Free Government Service: Crime Prevention," *Public Finance*, 19 (1964), pp. 383-392; and Lester C. Thurow, "Equity vs. Efficiency in Law Enforcement," *Public Policy*, 18 (Summer, 1970), pp. 451-62.

6. 157 Mich. 443, 473 (1909).

7. Wayne Lee Hoffman, "Equity and the Perceived Distribution of Urban Government Services," paper presented at the Annual Meeting of the American Political Science Association, Chicago, Illinois, September 2-5, 1976.

8. On community control, see Alan Altshuler, *Community Control* (New York: Pegasus, 1970); and Milton Kotler, *Neighborhood Government* (Indianapolis, Ind.: Bobbs-Merrill, 1969).

9. One excellent example of a "public-choice" approach to urban government is Robert L. Bish and Vincent Ostrom, *Understanding Urban Government*, (Washington, D.C.: American Enterprise Institute for Public Policy Research, 1973).

10. 44 U.S.L.W. 4789 (1976).

11. See Robert L. Lineberry, "Mandating Urban Equality: The Distribution of Municipal Public Services," *Texas Law Review*, 53 (December, 1974), pp. 26-59.

12. Annmarie H. Walsh, "Decentralization for Urban Management: Sorting the Wheat from the Chaff," in Willis D. Hawley and David Rogers, eds., *Improving the Quality of Urban Management* (Beverly Hills: Sage, 1974), p. 250.

13. Bish and Ostrom, *op. cit.*

14. E. S. Savas, "Municipal Monopolies versus Competition in Delivering Urban Services," in Willis D. Hawley and David Rogers, eds., *op. cit.*, ch. 15.

15. Edgar K. Browning, "How Much More Equality Can We Afford?" *The Public Interest*, 43 (Spring, 1976), p. 92.

BIBLIOGRAPHY

Equity, Equality, and Urban Policy: General

Harvey, David, *Social Justice and the City* (Baltimore: Johns Hopkins University Press, 1973).

Long, Norton, *The Unwalled City* (New York: Basic Books, 1972).

Miller, S. M., and Pamela Roby, *The Future of Inequality* (New York: Basic Books, 1970).

Rich, Richard C., "Institutional Arrangements and Equity in Urban Service Delivery," *Urban Affairs Quarterly,* 12 (March, 1977).

Shoup, Carl S., "Standards for Distributing a Free Government Service: Crime Prevention," *Public Finance,* 19 (1964), pp. 383-392.

Thurow, Lester C., "Equity vs. Efficiency in Law Enforcement," *Public Policy,* 18 (Summer, 1970), pp. 451-462.

The Politics and Economics of Urban Public Services

Antunes, George, and Kenneth Mladenka, "The Politics of Local Services and Service Distribution," in Louis H. Masotti and Robert L. Lineberry, eds., *The New Urban Politics* (Cambridge, Mass.: Ballinger, 1976), ch. 7.

Cox, Kevin R., *Conflict, Power, and Politics in the City: A Geographic View* (New York: McGraw-Hill, 1973).

Hirsch, Werner Z., "The Supply of Urban Services," in Harvey S. Perloff and Lowdon Wingo, Jr., eds., *Issues in Urban Economics* (Baltimore: Johns Hopkins University Press, 1968) pp. 477-525.

Jones, Bryan D., and Clifford Kaufman, "The Distribution of Urban Public Services," *Administration and Society,* 6 (November, 1974), pp. 337-360.

Murphy, Thomas P., and Charles R. Warren, eds., *Organizing Public Services in Metropolitan America* (Lexington, Mass.: D. C. Heath, 1974).

Tietz, Michael B., "Toward a Theory of Urban Public Facility Location," *Papers of the Regional Science Association,* 21 (1968), pp. 24-40.

Williams, Oliver P., *Metropolitan Political Analysis* (New York: Free Press, 1971).

Yates, Douglas, "Service Delivery and the Urban Political Order," in Willis D. Hawley and David Rogers, eds., *Improving the Quality of Urban Management* (Beverly Hills: Sage Publications, 1974), ch. 18.

Concepts, Methods, and Measurement

Alker, Hayward, and Bruce M. Russett, "On Measuring Inequality," *Behavioral Science,* 9 (July, 1964), pp. 207-218.

Hatry, Harry P., "Measuring the Quality of Public Services," in Willis D. Hawley and David Rogers, eds., *Improving the Quality of Urban Management* (Beverly Hills: Sage Publications, 1974), ch. 2.

———, and Diana R. Dunn, *Measuring the Effectiveness of Local Government Services: Recreation* (Washington, D.C. The Urban Institute, 1971).

Lineberry, Robert L., "Who is Getting What? Measuring Urban Service Outputs," *Public Management*, 58 (August, 1976), pp. 13-18.

———, and Robert E. Welch, Jr., "Who Gets What: Measuring the Distribution of Urban Services," *Social Science Quarterly,* 54 (March, 1974), pp. 700-712.

Ostrom, Elinor, "Exclusion, Choice, and Divisibility: Factors Affecting the Measurement of Urban Agency Output and Impact," *Social Science Quarterly,* 54 (March, 1974), pp. 691-699.

Ridley, Clarence E., and Herbert A. Simon, *Measuring Municipal Activities* (Chicago: International City Managers' Association, 1938).

The Urban Institute and the International City Management Association, *Measuring the Effectiveness of Basic Municipal Services* (Washington, D.C.: The Urban Institute, 1974).

The Law and Municipal Services

Abascal, Ralph S., "Municipal Services and Equal Protection: Variations on a Theme by *Griffin v. Illinois,*" *Hastings Law Journal,* 20 (May, 1969), pp. 1367-1391.

Bond, Kenneth W., "Toward Equal Delivery of Municipal Services in the Central Cities," *Fordham Urban Law Journal,* 4 (Winter, 1976), pp. 263-287.

Fessler, Daniel W., and Lucy S. Forrester, "The Case for the Immediate Environment," *Clearinghouse Review,* 4 (May and June, 1970).

———, and Charles M. Haar, "Beyond the Wrong Side of the Tracks," *Harvard Civil Rights—Civil Liberties Law Review,* 6 (May, 1971), pp. 442-465.

Graham, Robert L., and Jason H. Kravitt, "The Evolution of Equal Protection—Education, Municipal Services, and Wealth," *Harvard Civil Rights—Civil Liberties Law Review,* 7 (Jan., 1972), pp. 102-213.

Lineberry, Robert L., "Mandating Urban Equality: The Distribution Municipal Public Services," *Texas Law Review,* 53 (December, 1974), pp. 26-59.

Merget, Astrid E., and William M. Wolff, Jr., "The Law and Municipal Services: Implementing Equity," *Public Management,* 58 (August, 1976), pp. 2-8.

Note, "Equalization of Municipal Services: The Economics of *Shaw* and *Serrano,*" *Yale Law Journal,* 82 (November, 1972), pp. 89-123.

Ratner, Gershon M., "Inter-neighborhood Denials of Equal Protection in the Provision of Municipal Services," *Harvard Civil Rights—Civil Liberties Law Review,* 4 (Fall, 1968), pp. 1-64.

Tussman, Joseph, and Jacobus tenBroek, "The Equal Protection of the Laws," *California Law Review,* 37 (Sept., 1949), pp. 341-381.

Bureaucracies, Consumer Preferences, and the Distribution of Urban Services

Aberbach, Joel D., and Jack L. Walker, "The Attitudes of Blacks and Whites toward City Services: Implications for Public Policy," in John P. Crecine, ed., *Financing the Metropolis* (Beverly Hills: Sage Publications, 1970), ch. 18.

Bish, Robert L., and Vincent Ostrom, *Understanding Urban Government* (Washington, D.C.: American Enterprise Institute for Public Policy Research, 1973).

Davis, Kenneth C., *Discretionary Justice* (Baton Rouge: Lousiana State University Press, 1969).

Eisenger, Peter K., "The Pattern of Citizen Contacts with Urban Officials," in Harlan Hahn, ed., *People and Politics in Urban Society* (Beverly Hills: Sage Publications, 1972), ch. 2.

Fowler, Floyd J., Jr., *Citizen Attitudes Toward Local Government, Services and Taxes* (Cambridge, Mass.: Ballinger, 1974).

Hawley, Willis D., and David Rogers, eds., *Improving the Quality of Urban Management* (Beverly Hills: Sage Publications, 1974).

Hirschman, Albert O., *Exit, Voice, and Loyalty: Responses to Decline in Firms, Organizations, and States* (Cambridge, Mass.: Harvard University Press, 1970).

Hoffman, Wayne Lee, "Equity and the Perceived Distribution of Urban Government Services," paper presented at the Annual Meeting of the American Political Science Association, Chicago, Illinois, September 2-5, 1976.

Jacob, Herbert, "Contact with Government Agencies: A Preliminary Analysis of the Distribution of Government Services," *Midwest Journal of Political Science,* 16 (February, 1972), pp. 123-146.

Lipsky, Michael, "Toward a Theory of Street-Level Bureaucracy," in Willis D. Hawley, et al., *Theoretical Perspectives on Urban Politics* (Englewood Cliffs, N.J.: Prentice-Hall, 1976), ch. 9.

Mladenka, Kenneth R., "Citizen Demand and Bureaucratic Response: Direct Dialing Democracy in a Major American City," *Urban Affairs Quarterly,* 12 (March, 1977).

Savas, E. S., "Municipal Monopolies versus Competition in Delivering Urban Services," in Willis D. Hawley and David Rogers, eds., *Improving the Quality of Urban Management* (Beverly Hills: Sage Publications, 1974), ch 15.

———, "Municipal Monopoly, "*Harper's Magazine* (December, 1971), pp. 55-60.

Schuman, Howard, and Barry Gruenberg, "Dissatisfaction with City Services: Is Race an Important Factor?" in Harlan Hahn, ed., *People and Politics in Urban Society* (Beverly Hills: Sage Publications, 1972), ch. 14.

Sjoberg, Gideon, Richard A. Brymer, and Buford Ferris, "Bureaucracy and the Lower Class," *Sociology and Social Research*, 50 (April, 1966), pp. 325-337.

Stipak, Brain I., "Citizen Evaluations of Urban Services as Performance Indicators in Local Policy Analysis," unpublished Ph.D. dissertation, University of California at Los Angeles, 1976.

Tiebout, Charles M., "A Pure Theory of Local Expenditures," *Journal of Political Economy,* 64 (October, 1956). pp. 416-424.

**Empirical Studies of the Distribution
of Urban Taxes and Services**

American Library Association, *Access to Public Libraries* (Chicago: American Library Association, 1963).

Antunes, George, and John Plumlee, "The Distribution of an Urban Public Service: Ethnicity, Socioeconomic Status, and Bureaucracy as Determinants of the Quality of Neighborhood Streets," *Urban Affairs Quarterly*, 4 (March, 1977).

Blank, Blanche D., et al., "A Comparative Study of an Urban Bureaucracy," *Urban Affairs Quarterly,* 4 (March, 1969), pp. 343-354.

Bloch, Peter B., *Equality in the Distribution of Police Services: A Case Study of Washington, D.C.* (Washington, D.C.: The Urban Institute, 1974).

Boots, Andrew, et al., *Inequality in Local Government Services: A Case Study in Local Roads* (Washington, D.C.: The Urban Institute, 1972).

Chance, Truett L., "The Relation of Selected City Government Services to Socio-economic Characteristics of Census Tracts in San Antonio, Texas," unpublished Ph.D. dissertation, University of Texas at Austin, 1970.

Fisk, Donald, et al., *Measuring the Effectiveness of Local Government Recreation Services* (Washington, D.C.: The Urban Institute, 1972).

Fisk, Donald, and Cynthia A. Lauver, *Equality of the Distribution of Recreation Services: A Case Study of Washington, D.C.* (Washington, D.C.: The Urban Institute, 1974).

Gold, Steven D., "The Distribution of Urban Government Services in Theory and Practice: The Case of Recreation in Detroit," *Public Finance Quarterly,* 2 (January, 1973), pp. 107-130.

Jones, Bryan D., "Distributional Considerations in Models of Urban Government Service Provision," *Urban Affairs Quarterly,* 12 (March, 1977).

Kasperson, Roger, "Toward a Geography of Urban Politics: Chicago, a Case Study," *Economic Geography,* 11 (April, 1965), pp. 95-107.

Levy, Frank S., Arnold J. Meltsner, and Aaron Wildavsky, *Urban Outcomes: Schools, Streets, and Libraries* (Berkeley: University of California Press. 1974).

Lineberry, Robert L., "Equality, Public Policy and Public Services: The Underclass Hypothesis and the Limits to Equality," *Politics and Policy,* 4 (December, 1975), pp. 67-84.

Lyon, David, "Capital Spending and the Neighborhoods of Philadelphia," *Business Review* (of the Federal Reserve Bank of Philadelphia), (May, 1970), pp. 16-27.

Martin, Lowell A., *Library Response to Urban Change: A Study of the Chicago Public Library* (Chicago: American Library Association, 1969).

Mladenka, Kenneth, "Organizational Rules, Service Equity, and Distributional Decisions in Urban Politics: A Comparative Analysis," paper presented at the Annual Meeting of the American Political Science Association, Chicago, Illinois, September 2-5, 1976.

———, "Servicing the City: The Distribution of Municipal Services," unpublished Ph.D. dissertation, Department of Political Science, Rice University, 1975.

———, and Kim Q. Hill, "The Distribution of Urban Police Services," *Urban Affairs Quarterly,* 12 (1977).

Oldman, Oliver, and Henry Aaron, "Assessment–Sales Ratios under the Boston Property Tax," *National Tax Journal,* 18 (1965), pp. 36-49.

Pechman, Joseph A., and Benjamin A. Okner, *Who Bears the Tax Burden?* (Washington, D.C.: Brookings Institution, 1974).

Weicher, John C., "The Allocation of Police Protection by Income Class," *Urban Studies* 8 (October, 1971), pp. 207-220.

INDEX

Advisory Commission on Intergovernmental Relations, 88, 90.
Allbrandt, Roger, 117
American Library Association, 42, 120, 122, 186
Annexation, 44
Austin, Texas, 59-60, 63

Bastable, Charles F., 151
Beal v. *Lindsay*, 33
Bemis, Edward, 165-166
Berke, Joel, 89
Birmingham, Alabama, 120
Bish, Robert, 175, 195
Blacks, 75, 79, 101
  and library services, 120-121
  and police, 38-39
  in San Antonio, 54-55
  and taxes, 95
Boston, 91
Brown, Claude, 9
Brown v. *Board of Education,* 31
Buchanan, James, 73
Budgeting, 137, 151-152
  see also Incrementalism
Bureaucracies, 17-18, 145, 159, 174, 184, 191, 193-194
  decision rules, 12, 64-65, 146, 153-160
  discretion, 150-153
  as monopolies, 132, 163, 165-172, 175, 177, 195-196
  and public policy, 148-160

Chicago, 61, 121, 167, 186
Chicanos, 55, 60, 75, 79, 101, 113, 114
Coleman, James S., 31, 32, 69
Common law, 43-45

Community control, 191, 195
Compensation, 33-35, 37-38
Cox, Archibald, 25, 48
Cox, Kevin, 73
Crecine, John Patrick, 9, 64, 151

Dahl, Robert A., 79
Dye, Thomas R., 18

Ecological hypothesis, 62-64, 75, 134, 136-137, 146-148, 183
  and fire protection, 117-119
  and library services, 124-125
  and parks, 112-113
  and taxes, 97-98
Education, see Schools
Eisenger, Peter, 172
Elections, 160-161, 173-174
Elites, 61, 159
  see also power, political
Equal Protection clause, see Fourteenth amendment
Equality, 25-27, 186-187, 196-197
  idea of, 27-30
  input and output, 31-33, 36-37, 193
  of opportunity, 33-34
"Exit," 166-172, 174-176, 186, 196
Externalities, 73-74

Fire protection, 44, 77
  distribution of, 117-119
Fourteenth amendment, 26, 28, 45, 48, 192
France, Anatol, 181

Galbraith, John Kenneth, 25, 26
Glazer, Nathan, 181

ABOUT THE AUTHOR

ROBERT L. LINEBERRY is Professor of Political Science and Urban Affairs at Northwestern University. He taught at the University of Texas from 1967-1974. His articles have appeared in legal periodicals like *Georgetown Law Journal* and *Texas Law Review,* and in social science journals such as *American Political Science Review, American Sociological Review, Social Science Quarterly,* and *Journal of Politics.* With Ira Sharkansky, he authored *Urban Politics and Public Policy* (New York: Harper & Row, 2nd ed., 1974); and with Louis H. Masotti, he edited *The New Urban Politics* (Cambridge, Mass.: Ballinger, 1976).

The work of many prominent social scientists—in writing,
research, analysis and synthesis—is of interest and
importance far beyond the boundaries of traditional
disciplines and sub-fields. It is to present such
works—diverse in subject and audience—that the
SAGE LIBRARY OF SOCIAL RESEARCH
exists. Volumes in this series are published
originally in paper; clothbound library editions
are also available.

1365

SAGE PUBLICATIONS
The Publishers of Professional Social Science
Beverly Hills ● London

DATE DUE